Concealing Coloration in Animals

Concealing Coloration in Animals

Judy Diamond and Alan B. Bond

The Belknap Press of
Harvard University Press
Cambridge, Massachusetts
London, England
2013

Printed in the United States of America

Library of Congress Cataloging-in-Publication Data

Diamond, Judy.
Concealing coloration in animals / Judy Diamond and Alan B. Bond.
pages cm
Includes bibliographical references and index.
ISBN 978-0-674-05235-2
1. Protective coloration (Biology) I. Bond, Alan B., 1946- II. Title.
QL767.D53 2013
581.4'7--dc23

2012041730

For Rachel and Benjamin

Contents

Preface

CONCEALING COLORATION FUSES animals with their natural world, making them as bound to conceal themselves as they are to breathe. Such remarkable creatures amaze and delight human observers, but understanding their evolution requires a process grounded more in science than in nature study. Does the resemblance of an animal to its surroundings actually increase its chances of survival? And if the resemblance has benefits, why don't all animals have concealing coloration? These are old questions, in some cases dating back to classical antiquity, and they are deeply rooted in the narrative of biological thought.

Throughout this lengthy history, a few individuals stand out. Alfred Russel Wallace and his better-known colleague Charles Darwin discovered natural selection, the driving mechanism of evolution. It was Wallace, not Darwin, who laid the basis for a modern interpretation of concealing coloration, applying evolutionary principles to the full range of animal color patterns. The publication of our book marks one hundred years since Wallace's death on November 7, 1913, at the age of ninety.

The Dutch biologist Nikolaas "Niko" Tinbergen shaped our understanding of animal coloration in an entirely different way. As a professor at Oxford University, Tinbergen inspired generations of scientists to investigate the role of evolution in animal behavior. Tinbergen's students went on to conduct many of the classic experiments on concealing coloration that are described in this book. In 1973, Tinbergen shared the Nobel Prize in Physiology or Medicine with Karl von Frisch and Konrad Lorenz for work on the study of social behavior and instinct. Just as Wallace provided the theoretical framework for understanding color evolution,

Tinbergen advocated the experimental approaches that would be used to test and verify those theories.

This book is an introduction to ideas that span the life sciences. In each chapter, we chose a set of organisms, researchers, and experimental studies that exemplify principles of concealing coloration. Animal color patterns are one of the most approachable aspects of evolution. The role of color in concealment beautifully illustrates a scientific triad—the interdependence of psychology, genetics, and ecology in shaping the features of species. Animals that resemble twigs, leaves, stones, and moss appear at first glance to be miraculous, uncanny creations of visual perfection and diversity. Scientific experiments explain how these phenomena arise through natural evolutionary processes, perhaps less mysterious but no less astonishing.

Concealing Coloration in Animals

Part One

CONCEALMENT

Now it is remarkable in how many cases nature gives this boon to the animal, by colouring it with such tints as may best serve to enable it to escape from its enemies or to entrap its prey.

ALFRED RUSSEL WALLACE, "Mimicry, and Other Protective Resemblances Among Animals"

The lantern bug (*Pyrops candelaria*) from Borneo has evolved a color pattern that makes it stand out from the background. Can you find the locust on the same log on the lower right? The spiky grouse locust (*Discotettix* sp.) looks like a tuft of moss. Photograph by Judy Diamond.

1. Disappearing Acts

WALK THROUGH A PATCH of prairie or along a beach just below the tide line; hike through the sand and rocks of a desert wash or among giant rainforest trees. Even as a seasoned observer, you may not see much sign of animal life. The creatures around you will be most noticeable when they chirp, twitter, or twitch; when they abruptly dive into the underbrush; or when their wings flash during takeoff. As a general rule, animals tend to blend in, rather than stand out, merging into the hues and patterns of the background. Small insects and birds of the forest canopy become a blur amid the yellow-green foliage; those that forage in the dim light on the ground below are abstract compositions in dark reddish brown. The relationship is strongest where the surroundings are most uniform. Many mammals and birds of the far north are white, at least in winter; fishes of the open ocean look dark blue from above and silvery from below; and desert birds and lizards match the colors of their local soils.

An animal's coloration is a product of its evolutionary history. When a color pattern increases the odds that an individual will survive and reproduce, it gradually becomes more abundant in the population. *Population* commonly refers to the number of people that live in some geographic region, but to biologists, a population consists of the members of the same species that live in a single area and interbreed. Features that aid survival become more frequent in the population as they are passed on to subsequent generations. There are many ways that color helps an animal to survive. It can, for example, assist in maintaining an animal's body temperature or protect against ultraviolet radiation. Color is a central element in social communication, enabling animals to

The cactus wren (*Campylorhynchus brunneicapillus*) shows a general resemblance to the colors of the Mohave Desert. Photograph by Benjamin Bond.

recognize members of their own species, attract mates, and signal status or motivation. The color patterns of some animals warn other species that they are dangerous or poisonous. But one of the most pervasive functions of color is concealment, where color makes prey animals difficult to find or helps predators to hide from their prey.

Invisibility is a staple of science fiction, but in the real world, color concealment is at best an imperfect strategy. Most environments are a complex patchwork of splotches of different colors and textures, so no single color pattern is invariably concealing. A moth that is perfectly hidden when it settles on tree bark will be blatantly obvious on a nearby house or wooden fence. A pale green grasshopper invisible among the stems of prairie grasses becomes an instant target for hungry meadowlarks when it lands

The differential grasshopper (*Melanoplus differentialis*) from Montana is shown on smooth brome grass (*Bromus inermis*). Photograph by Alan Bond.

ABOVE: The eyespot on the owl butterfly (*Caligo memnon*) from Costa Rica can startle predators, increasing the odds that the butterfly will escape an attack. Notice the bite taken out of the hind wing by a bird that missed its target. Photograph by Judy Diamond. **BELOW:** Looking like another, more dangerous animal, such as a bumblebee, can help this snowberry clearwing (*Hemaris diffinis*) avoid predators. Photograph by Alan Bond.

on bare ground. For these animals, survival depends on where in the environment they choose to rest and how long they can remain motionless. Species that travel through multiple, visually distinctive environments may evolve an average appearance that is not perfectly concealing in any location. For both prey and predators, the survival value of a color pattern depends on how well it misleads. Walking sticks look amazingly like twigs, and they readily fool young, inexperienced predators. But twigs do not have legs that angle out forward and backward from the stem, and eventually, more experienced predators learn to distinguish the bugs from the branches.

Amid the confusion and clutter of the natural environment, predators concentrate their search on such telltale signs, ignoring everything else. There is a great benefit to this: When you specialize in searching for specific details, even cryptically colored prey can seem obvious. But there is also a cost to paying too close attention, since you can become blind to the alternatives. When a bird searches intently for caterpillars that look like twigs, it misses nearby moths that look like bark. When a cat is momentarily distracted by a flash of color or a quick movement among the branches above, it walks right by the mouse sitting motionless on the ground. The benefit of concealing coloration is not that it provides an ironclad guarantee of survival, but that it consistently yields a small edge in the chance of living through each successive threatening encounter. At a minimum, even a tiny delay between the approach of a predator and its subsequent attack can help a prey animal escape. And at best, the prey will be completely overlooked.

The evolution of concealment is different from that of color patterns that serve to advertise species identity, social status, or inedibility. Advertising coloration is maximally visible to the intended recipient of the message; it is a signal that announces the presence of the animal and broadcasts a particular set of messages: "I am a red-winged blackbird," or "I am sexy," or "I am poisonous." Concealing coloration accomplishes the opposite; it reduces the chance of detection. As a result, the evolution of concealing coloration is almost entirely a product of the visual sensitivity, perceptual mechanisms, and cognitive capabilities of the animal's most important predators or their most favored prey. Studies of concealing coloration provide some of the clearest examples of how the evolution of a feature in one organism can be driven by the psychology of another.

Concealment is not always about hiding from sight. There are concealment strategies for almost every sensory system. Animals silence their mating songs when predators are near, or they change the pitch of their calls so they cannot be easily located. Some animals mask their scent to hide from predators, and others minimize the vibrations produced when they move. Concealment through color and pattern isn't the only way that animals protect themselves, but it has been more extensively investigated than any other strategy.[1]

A Saharan Epiphany

The sand dunes of the Sahara Desert produced an unlikely proponent of the evolution of concealing coloration. To satisfy a passion for natural history, an Anglican priest named Henry Baker Tristram spent his winters exploring the Middle East and North Africa. During the 1850s, he endured months in the saddle conducting biological field surveys, hauling his bird skins and plant presses on luggage camels through the most remote parts of Africa. He bargained with locals for food and water, dodged an attack by wild stallions, engaged in an armed standoff with desert bandits, and lost pack animals to hyenas, and he wrote extensively about the resemblance between the colors of desert animals and their surroundings.

Tristram was most interested in the permanent residents of the dune fields and rocky washes of the Sahara—ground birds such as desert larks and sandgrouse, as well as small mammals, snakes, and lizards. He was impressed by the fact that all these unrelated creatures displayed much the same uniform, sandy coloration, and he argued that it must have resulted from a common evolutionary process. Tristram noted that there were individual differences in color among all animals, and he reasoned that if a particular color increased the animal's resemblance to its surroundings, it would provide a slight increase in the chance of escaping from natural enemies, thereby giving that animal a decided evolutionary advantage over other forms of the same species. This would result in making the animals with that special color more abundant.[2]

The description Tristram provided was a lucid summary of the modern concept of evolutionary adaptation. Within a single species, individuals vary from one to another in many features or traits, such as the amount or kind of pigment in their fur or feathers. These traits are to some degree heritable, being passed on from parents to offspring. Some traits provide a survival advantage to the individuals. Among the animals of the deep Sahara, individuals that more closely match the rust red, golden brown, or pale gray of the desert soils are somewhat more likely to survive, reproduce, and contribute genes to later generations. Over many generations, this process results in populations of animals with colors that resemble their desert habitat.

Tristram's evolutionary insight was inspired by a paper written in 1858 by a thirty-five-year-old naturalist named Alfred Russel Wallace. The essay appeared in the *Proceedings of the Linnean Society of London,* along with excerpts of two of Charles Darwin's previously unpublished writings. Tristram saw in these short essays a powerful, unifying explanation for what he had repeatedly observed in the desert but had not previously understood. A similar argument for natural selection was developed in over five hundred closely reasoned pages, employing a huge variety of examples, in Charles Darwin's *On the Origin of Species,* which was published the following year. But Tristram's paper refers only to Darwin's and Wallace's earlier sparse outline of the argument. The *Origin* was not at the time available in Algeria.[3]

Wild about Butterflies

In the first edition of the *Origin,* Darwin mentioned concealing coloration only in passing, calling it an adaptation of trifling importance. After all, although grouse often resemble the color of their surroundings, they are nonetheless killed by visual predators, particularly goshawks, suggesting that their apparently concealing coloration was far from a perfect protection. It was really only after Darwin had read Henry W. Bates's work on butterfly mimicry that he became a confirmed color enthusiast. Bates had made a joint expedition up the Amazon with his close friend Alfred Russel Wallace in 1848 while they were both still in their twenties. For two young naturalists of limited means, it was a great adventure that they had intended to fund by selling duplicates of the specimens they gathered. Wallace returned to England in 1852, losing his collections in a fire at sea. But from his notebooks, he was able to publish a variety of papers and several books on his travels. He returned to the field two years later to explore the jungles of Malaysia and Indonesia.

Bates remained in Brazil for another seven years, accumulating a collection of over 14,000 insects, of which 8,000 were new to science. Bates was particularly struck by the Heliconid butterflies, which are diverse, brightly colored, and very conspicuous in the Amazonian jungle. He noted that a number of species from an unrelated butterfly family, the Pieridae, had color patterns that looked amazingly like Heliconids, deviating from the plain yellow or white that is more typical of their group. Bates determined that the colorful Heliconids, which he termed *models*, were acutely distasteful to birds, inducing nausea and vomiting. The Heliconid-patterned Pierids, which lacked the offensive chemicals that would make birds avoid them, were *mimics*, in Bates's terminology. Bates reasoned that the resemblance of the mimics to the models must have been the result of natural selection. Predators that had learned from their unfortunate experience with the models would tend to avoid anything that even slightly resembled such offensive butterflies, thereby promoting the evolution of mimics.

What has come to be called Batesian mimicry was a significant piece of evidence in favor of Darwin's theory. Bates published his description of the mimetic Amazonian butterflies in 1862, and Darwin was unreservedly enthusiastic. He wrote to Bates, "I have just finished after several reads your Paper. In my opinion it is one of the most remarkable & admirable papers I ever read in my life. The mimetic cases are truly marvellous & you connect excellently a host of analogous facts. The illustrations are beautiful & seem very well chosen . . . I rejoice that I passed over whole subject in the Origin, for I shd. have made a precious

The six species of butterfly and two species of moth shown above are all members of a mimicry ring. The butterfly in the center, the tiger-striped longwing (*Heliconius ismenius clarescens*), is considered the original model. All the species are from Barro Colorado Island, Panama. Photograph by Christian Ziegler (Getty Images).

mess of it. You have most clearly stated & solved a wonderful problem" (Darwin 1862, 1).[4]

Godfather of Animal Coloration

Unlike Darwin, who was born to wealth, Wallace was short of funds and scrabbling for an income for most of his life. With no money to attend the university, his formal education ended at fourteen. But Wallace was nevertheless willing to take intellectual risks. He came independently to his understanding of natural selection much the way Darwin did—in a moment of illumination after an arduous and stimulating period of field research. Darwin ruminated and wrote privately about his ideas for more than twenty years. Wallace responded to his insight by dashing off a paper from the island of Ternate in eastern Indonesia while recovering from malaria, and he mailed it to Darwin in all innocence, asking for his comments.

When Darwin received Wallace's manuscript, he was stunned to discover that the idea he had been mulling over for decades and was just on the verge of publishing had occurred to someone else. On June 18, 1858, he sent off Wallace's manuscript to his friend, Charles Lyell, with a few brief, distressed comments: "I never saw a more striking coincidence. if Wallace had my M.S. sketch written out in 1842 he could not have made a better short abstract! Even his terms now stand as Heads of my Chapters" (Darwin 1858). Lyell and Joseph Hooker arranged a delicate compromise, in which Wallace's paper was published along with excerpts from letters that Darwin had previously sent concerning the mechanism of natural selection. It was this publication that Tristram read in the Sahara.[5]

Wallace was far more receptive to unconventional notions than Darwin. He was an advocate for social justice and land reform and a believer, much to the distress of his colleagues, in spiritualism and other psychic phenomena. Wallace was also a careful and patient observer, and he reasoned synthetically from evidence with brilliance and originality. From very early on, he was a resolute proponent of the evolutionary significance of concealing coloration. To generations of naturalists, from Tristram to the present day, Wallace was the godfather of coloration research.

After returning to England, Wallace mined the existing literature, as well as his own experiences in Asia, to produce an extensive list of examples of concealing coloration in both vertebrates and invertebrates. In 1867, he published a long review article, entitled "Mimicry, and Other Protective Resemblances Among

Animals," in which he laid the foundations for understanding the evolution of animal coloration. Wallace began his argument by remarking on the general resemblance between the colors of animals and those of their backgrounds: "Arctic animals are white, desert animals are sand-coloured, dwellers among leaves and grass are green, nocturnal animals are dusky. These colours are not universal, but are very general, and are seldom reversed" (Wallace 1867, 40). He then noted that in many cases the resemblance goes further, yielding an exact replica of leaves, sticks, mossy twigs, or flowers. These animals often show specialized behaviors that aid in the deception and render the concealment more natural.

Wallace envisioned a gradation of degrees of concealment, from simple correspondence between the color of the animal and that of the background to the incorporation of characteristic pattern elements to a mimetic resemblance to common inedible objects in the environment. Finally, he came to discuss true Batesian mimicry, the evolved resemblance between a palatable mimic and an aversive, warningly colored model. By tying these phenomena together, Wallace argued for a continuity between mimicry and other forms of protective coloration. Mimicry was at one extreme of a range of adaptive mechanisms that together explained the evolution of color resemblance. In this fashion, he argued that animal coloration was a primary illustration of the power of natural selection.[6]

Darwin evidently found Wallace's and Bates's arguments compelling. In the sixth edition of the *Origin* in 1872, he gave a detailed account of Bates's ideas on insect mimicry and generalized to the importance of concealing coloration, arguing for the selective value of even very small increments in background resemblance. Evolutionary explanations of the origin of animal coloration met with a surprising amount of resistance, however.

Some people objected to the idea of evolution on moral or religious grounds, but there were many others who could accept the notion of natural selection but could not see how it could possibly account for concealing coloration. Even among related species, color patterns were so diverse and their protective effectiveness was so clearly limited that an evolutionary explanation seemed unnecessary. The strongest and most articulate objections came from the twenty-sixth president of the United States.[7]

The Naturalist President

Theodore Roosevelt doesn't elicit the same universal reverence among Americans as Washington or Lincoln, but his trademark features—the Rough Rider uniform, the bottle-bottom glasses, the walrus moustache, the toothy grin—make him one of the most readily recognized of America's chief executives. Aside from the better-known accomplishments of his administration, Roosevelt had a distinction rarely seen among American politicians: He published repeatedly in peer-reviewed scientific journals. It is impressive, as Stephen Jay Gould noted, that in the midst of the political turmoil leading up to the bizarre, four-way election of 1912, Roosevelt should have taken the time to produce a 112-page review (and diatribe) titled "Revealing and Concealing Coloration in Birds and Mammals" for the *Bulletin of the American Museum of Natural History.*

Natural history was Teddy Roosevelt's first and deepest passion. It provided him with the primary focus of his youth, his extensive personal museum with mounted specimens, his field notebooks and diaries. It drove him to seek the challenges of exploring wild places, facing physical demands that he later credited with transforming him from a frail and asthmatic child into a vigorous and combative adult, and it underlay his emotional attachment to the conservation of wilderness. During his presidency, hundreds of wildlife reserves, national forests, national parks, and monuments received federal protection. Even when actively engaged in presidential politics, Roosevelt's naturalist instincts were unabated. His sister Corinne writes of walking with him on the White House grounds, when Roosevelt abruptly stopped and picked up a tiny feather from the path. "Very early for a fox sparrow," he muttered. Corinne asked how he could possibly know that the feather came from a fox sparrow, and Roosevelt replied, "Well, you see I have really made a great study of sparrows" (Robinson 1921, 232). Given his personal history, one might expect Roosevelt to have adopted a rigorously Darwinian approach to coloration. He was, after all, writing well after the discovery of natural selection, and he was so familiar with the *Origin* that he carried a leather-bound copy in his saddlebags.[8]

Roosevelt considered himself a preeminent authority on common sense in natural history, and this overrode his attachment for evolution. From very early on, he was outspokenly skeptical of adaptive interpretations of animal color patterns because they seemed to violate his own direct experience. There are two related premises embedded in Roosevelt's skepticism, two commonsense views that he found difficult to reconcile with natural selection. The first relates to the diversity of color strategies among related animals. If concealing coloration provides an evolutionary advantage to fox sparrows, for example, why do white-crowned and white-throated sparrows, which feed in similar circumstances, carry such conspicuous patches and streaks? Why have indigo buntings, eastern towhees, and rose-breasted grosbeaks not evolved a similarly cryptic coloration? If concealing coloration was selectively advantageous for one of these species, why not for all of them?

The diversity of coloration was a serious concern, but it was the second premise that was Roosevelt's primary obsession. He simply could not see how color resemblance could have any meaningful effect on survival since even the most cryptically patterned species were only truly hidden when they were motionless. No animal with a more demanding lifestyle than an aphid or a barnacle could remain motionless forever, and motion instantly attracts visual attention. It is literally impossible for visual predators to overlook moving prey. As soon as an animal begins to move, it becomes immediately obvious, irrespective of its coloration.

Roosevelt argued that the behaviors used to evade capture are more important in ensuring survival of a prey animal than its degree of resemblance to the background. On the other hand, any animal, even a relatively conspicuous one, may be difficult to see some times and under some conditions. Roosevelt noted that even elephants are hard to see in dim light or in heavy underbrush (at least, he found them so). On the whole, Roosevelt argued, coloration had so little effect on an animal's survival that it was virtually irrelevant.[9]

Roosevelt's concerns were vividly illustrated in a study made in 1899 of the preserved stomach contents of American birds. Sylvester Judd, a biologist in the U.S. Department of Agriculture, displayed almost superhuman dedication in examining, over the course of four years, the last meals of some 15,000 birds. He classified all insect remains that he found. Judd cited Wallace as stating that all orthopteran insects (including grasshoppers, locusts, crickets, and mantids) are protectively colored. But his stomach content survey found that grasshoppers are eaten in massive numbers by birds, amounting to almost a fifth of the diet of catbirds, a quarter of that of house wrens, and 40 percent of the diet of meadowlarks.[10]

Many of these prey species were nearly invisible to human eyes in their customary habitats. But Judd later conducted behavioral experiments showing that sparrows and mockingbirds had no difficulty finding grasshoppers or katydids even when they were completely immobile and even when humans were utterly unable to see them. He found similar patterns of massive consumption of camouflaged caterpillars, weevils, and terrestrial spiders. Judd even discovered full-grown walking sticks, which are among the most cryptically colored insects in North America, in the stomachs of common grackles and cuckoos. Although he was sympathetic to evolutionary arguments, Judd's final inference was unequivocal: Cryptic coloration was of limited value in enabling animals to escape from their predators. As the Scottish naturalist Sir John Arthur Thomson pithily observed, "The desert insect does not escape the desert lizard, and the green insect on the twig is unhesitatingly picked off by the sharp-eyed bird who has made that its business" (Thomson 1914, 541).

One is compelled, therefore, to wonder: Is there a reason to expect all animals to evolve cryptic coloration? Is it even possible to attain a color pattern that is completely indistinguishable from the background? And if the resemblance is only approximate, how much protection does a prey animal or predator gain from it? Under what circumstances does the strategy of concealment work, and why does it so often seem to fail? Whose arguments turn out to be most compelling: Alfred Russel Wallace or the American president? How does a century of modern evolutionary research bring new insight into the surprisingly diverse mechanisms that underlie concealing coloration?

2. Mistaken Identity

SOME OF THE MOST unusual animals in the ocean live just below the surface in gigantic, entangled mats of free-floating kelp called sargassum. These creatures appear exactly like pieces of the intertwined stems, floating bladders, and fronds that make up their habitat. They look very little like their relatives from closer to shore. Drawn together by the convergence of ocean currents, this underwater forest covers a highly variable area, depending on time of year and prevailing winds. In some places, the forest is scattered in patches; in others, it is so thick it can slow sailing vessels. Sargassum forests are found in the Indian Ocean and in the Western Pacific, but by far the largest one is in the North Atlantic. It is called the Sargasso Sea.[1]

An early explorer of the Sargasso Sea was Peter Osbeck, a clergyman and botanist. Aboard the merchant vessel *Prince Charles*, Osbeck set sail from Sweden in 1750, bound for China and the East Indies. Osbeck was a disciple of Carl Linnaeus, the founder of biological classification, and his primary assignment was scientific exploration. Linnaeus had encouraged the young man to make the trip and had recommended him for the position. In return, Osbeck lavished attention on the natural history of the regions he passed through, bringing back numerous specimens of oriental plants and ultimately producing a detailed narrative of the voyage.

On the trip home, while his ship passed through the eastern edge of the Sargasso Sea, Osbeck eagerly trolled his nets through the mats of seaweed. He collected, among other exotic creatures, several specimens of the sargassum fish, a small and bizarrely colored member of the frogfish family. Osbeck was surprised to

Glauert's sea-dragons (*Phycodurus eques*) are related to seahorses. They live among floating mats of sargassum in the Indian Ocean and off the coast of Australia. Photograph by David Fleetham.

see that the fish's whole body was covered with small projections. He was, after all, a botanist, so he likened the projections to little sargassum leaves and went on to speculate, "Perhaps Providence has clothed this fish with *fulcra* resembling leaves, that the fishes of prey might mistake it for sea-weed, and not entirely destroy the breed" (Osbeck 1771, 113). Osbeck's note on this unusual fish was one of the first published accounts of object resemblance, a form of protective coloration in which an animal adopts the detailed appearance of an inedible object.[2]

Open-water predators such as dolphin fish and yellowtail amberjacks forage around the mats, and the sargassum fish's resemblance to a wad of seaweed may help it to avoid detection. But sargassum fish have additional means of escape: Both their pectoral and their pelvic fins are shaped more like hands than paddles, allowing them to climb deep into the tangles of the mat.

The sargassum fish (*Histrio histrio*) is protected by its resemblance to seaweed. The pale, rounded structures are the bladders that keep the sargassum afloat. Photograph by Oxford Scientific (Getty Images).

In addition, like all frogfishes, their gill openings are restricted to small circular holes just behind the pectoral fins. When under attack, they can rapidly expel water through these holes and zip away by jet propulsion.

Escaping predation is not the only benefit that sargassum fish derive from their leafy appendages, and it may not even be the primary one. Frogfish as a group are themselves highly effective predators. They sit and wait for an oblivious fish or crustacean to wander by, and then abruptly open their gigantic mouths, sucking in the doomed victim. Their feeding technique is remarkable for its speed and power. A frogfish can expand its mouth cavity twelvefold and gulp down prey larger than itself in milliseconds. All frogfish have evolved strikingly cryptic color patterns, often including protruding spines or wartlike blobs. So although the sargassum fish is the only frogfish that lives among floating seaweed, it is both predator and prey in the mats, and it uses much the same lurking behavior and coloration strategies as other members of its family.

Persistent floating mats of seaweed are a unique habitat, and the animal species that make up the self-sustaining community in the sargassum have a variety of features to cope with its demands. Many fishes use the mats as a nursery, where their young can shelter and feed until they are large enough to move into the open ocean. But the Sargasso Sea is also home to an assortment of animals that live nowhere else. These species are generally much smaller than their relatives along the continental coasts, they cling tightly to their weedy refuges, and, most of all, their coloration and pattern is strikingly similar to sargassum, often including leaflike appendages.

Two species of shrimp demonstrate the range of sargassum adaptations. Young slender sargassum shrimp strongly resemble leaflets of seaweed. They vary in color from yellowish to dark brown, exactly matching the range of colors of the sargassum fronds. Cerulean sargassum shrimp are globe shaped, and the young match the size and coloration of the sargassum air bladders, the spherical structures that keep the seaweed afloat. Both species show similarity to pieces of seaweed only when they are immature. When they grow larger, they no longer resemble leaves or bladders. Instead, they show complex mottled patterns that incorporate the full range of colors of the surrounding mats.[3]

Osbeck's descriptions of the sargassum animals helped scientists appreciate the diversity of life in these floating habitats. But it would fall to later generations to begin to explain how the adaptations work. Alexander Agassiz, a curator at the Museum of Comparative Zoology at Harvard, supervised another biological survey of the Sargasso Sea in 1877. He noticed an unusual aspect of the behavior of the animals that live in the mats. When a piece of sargassum is placed gently in a glass dish, it seems to be empty of life. But if the dish is then shaken, hundreds of tiny creatures emerge and scurry around in all directions, "eager to return to the particular spot best adapted to conceal them" (Agassiz 1888, 212–13). Agassiz suggested that when they are shaken loose, the animals return not just to some arbitrary shelter among the fronds of the plant, but to positions that are "best adapted to conceal them." This would require that the animals perceive the variations in the appearance of the plant and actively select specific locations and orientations that maximize the effect of their camouflage. It is a wonderful hypothesis, but it required an experimental test.[4]

Choosy Shrimp

Scientific progress in understanding sargassum animals seems to proceed by hundred-year increments. A century after Agassiz's survey and two centuries after Osbeck's discovery, two biologists from Woods Hole Oceanographic Institution used modern experiments to determine whether Agassiz was correct. Sally Hacker and Laurence Madin in 1991 tested Agassiz's assertion using two species of shrimp that live exclusively in sargassum mats.

Hacker and Madin traveled in an oceanographic research vessel to an area southeast of Bermuda, where they collected clumps of sargassum that contained the two shrimp species. Groups of shrimp were placed in cylindrical tanks on the deck of the ship, and in each tank, the shrimp were given choices between two types of habitat of equal size.

There were four experiments, each addressing a different hypothesis about shrimp behavior. The first study compared sargassum with a yellow plastic aquarium plant. The researchers asked whether

The slender sargasssum shrimp (*Latreutes fucorum*) resembles the fronds of sargassum. Photograph by Oxford Scientific (Getty Images).

there was something special about sargassum, perhaps some chemical or surface texture that made the seaweed more attractive than other substrates. Hacker and Madin found that neither species of shrimp was more attracted to sargassum than to plastic plants of a similar color, suggesting that there was nothing special about sargassum, as such.

The second experiment used two plastic plants, one yellow and one brown, to see whether the shrimp used color to discriminate between possible habitats. The slender sargassum shrimp, the species that most resembles fronds, proved to be more likely to choose the yellow than the brown plastic plant, showing that they used color to decide on a resting habitat. But color turned out not to be relevant for the cerulean shrimp, the species that resembled bladders. They did not distinguish between the two colors of plastic plant.

The third experiment used two pieces of sargassum that had parts removed: One piece had only the air bladders and the other had only fronds. This tested whether the shrimp discriminated between the shapes and sizes of different seaweed features. When the animals were allowed to choose between sargassum with fronds or with bladders, the slender shrimp overwhelmingly preferred to attach to branches with large numbers of fronds, and they aligned themselves along a frond when clinging to the plant. The cerulean shrimp absolutely preferred branches with lots of bladders, and they clung perpendicularly to the stem of the plant, corresponding to the usual orientation of the bladders.

In the fourth experiment, two different sizes of slender sargassum shrimp were placed in a tank with two plastic plants, one had large fronds and the other had small ones. The scientists observed that small slender shrimp preferred the plastic plant with the smaller fronds, but the larger shrimp did not discriminate between the sizes of fronds. In this fourth study, there was an additional experimental treatment. Half of these frond-size tests were in a tank with only the two plants, while the others included a sargassum fish in the tank with the shrimp. To prevent the experimental subjects from being eaten, the fish was confined to a ventilated transparent container within the tank. These tests showed that the presence of the predator made no difference to either shrimp in their choice of location.

The experiments suggested that Agassiz was correct, at least with respect to these two shrimp. Both species chose to attach themselves to substrates that best provided concealment. The two shrimp are similar in shape and coloration to particular natural objects, and when given a choice, they chose appropriate, matching habitats and oriented themselves in ways that enhanced the illusion. Like children playing hide-and-seek, the shrimp used multiple criteria in deciding where to hide. The shape and size of the plant part was most important, but the slender shrimp also recognized the color of the fronds and selected the plant that was closest to their own coloration. When they were not given a satisfactory choice, as when the cerulean shrimp had to choose between two leafy plastic plants, they immediately attached themselves to one of the alternatives anyway. It is apparently better to be clinging to something, even something unsuitable, than to be swimming around in the open water where you can be inhaled by a nearby predator. The shrimp focused on the form and color of the plant parts, rather than on the looming sargassum fish. Since the predator itself strongly resembles a mass of sargassum, perhaps the shrimp did not realize it was there. Or perhaps they were so desperate to get attached to the seaweed that even the presence of a predator could not make them move any faster.[5]

Evolution of Object Resemblance

The correspondence of the shrimp in shape, size, and coloration to an inedible object, combined with a strong tendency to choose particular matching environments and orientations, is diagnostic of object resemblance. The phenomenon is far from being limited to the fauna of the Sargasso Sea. Animals in many areas of the world resemble common inedible items in their environment, including stones, leaves, sticks, lichen, blocks of coral, and bird droppings. Species of fish, toads, butterflies, moths, praying mantids, and grasshoppers have all evolved resemblances to leaves, and the strategy is pervasive among leaf insects and tropical katydids. Object resemblance seems more common in tropical forests and marine environments, but that may just reflect

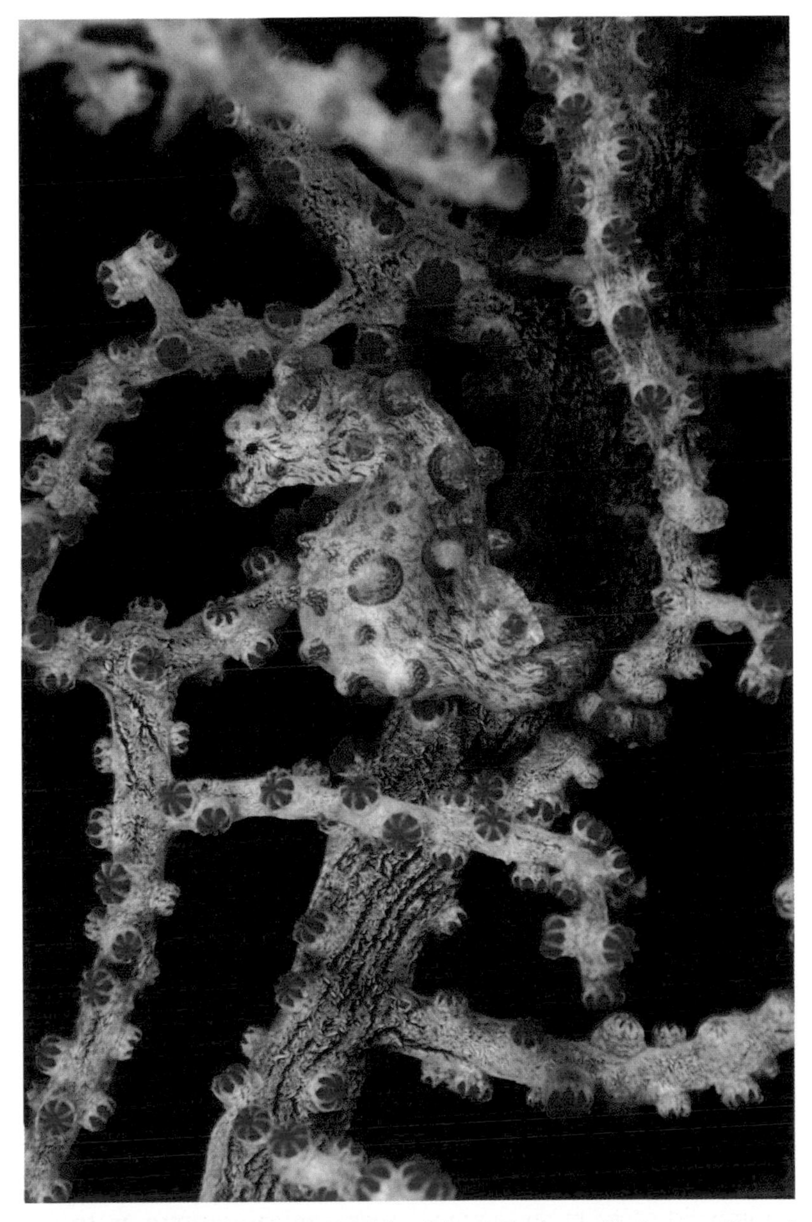

This pygmy seahorse (*Hippocampus bargibanti*) from Indonesia closely resembles the surrounding coral branches. Photograph by Wolfgang Poelzer (Getty Images).

The leaf-mimic katydid (*Typophyllum* sp.) from the Amazon rainforest in Peru looks like a dead leaf down to the veins and patches of discoloration. Photograph by Martin Shields (Getty Images).

the huge diversity of species that occur there. There are certainly stick-mimic caterpillars and moths that resemble leaves or bird droppings even in temperate regions.[6]

Object resemblance provides some of the most unmistakable illustrations of the role of natural selection in concealing coloration. The explanation for the gradual evolution of object resemblance begins with the traits of a given species that are naturally variable, such as the yellow or brown coloration of the slender sargassum shrimp. In addition to coloration, traits may include variations in size, growth rate, speed of movement, and even behaviors that lead to preferences in where an animal likes to rest or feed. No two shrimp or fish or dogs or people are exactly alike. These variations can result in differences in how the organisms respond to their environment, and to the degree that the variability is heritable, the traits are passed on from one generation to the next.

Depending on environmental conditions, even subtle differences can influence which individuals survive long enough to reproduce. For example, young sargassum mats tend to be lighter colored than older mats. In the young mats, yellow shrimp would be harder for predators to find than brown shrimp. This doesn't mean that all brown shrimp will be lunch and all yellow ones will be overlooked. Evolution is a matter of relative mortality. All that is necessary is that on average, a yellow shrimp will survive somewhat longer than a brown one, giving yellow shrimp a greater chance to grow up and breed. Over successive generations, yellow shrimp will become more common in the population.

Sargassum animals illustrate two characteristics of concealing coloration. The first is that object resemblance is not just a matter of color and pattern matching. Alfred Russel Wallace remarked that, in many cases, it is aided by specific behavior patterns that enhance the illusion. Animals that are attracted to the appropriate type of sargassum branch or that orient correctly with respect to the stem survive longer than others that do not display those behaviors. Individuals pass on the genes that code for both color and behavior to their offspring, and ultimately those traits will come to dominate in the population.

The second characteristic of object resemblance is that it may only work at a particular stage in an animal's life cycle. For example, when sargassum shrimp of either species are small, they are uniformly colored and bear a strong similarity to particular parts of the plant. When they grow larger, they can no longer effectively imitate the fronds or the bladders of the sargassum. Larger shrimp of both species show complex mottled patterns that incorporate the full range of colors of the surrounding kelp mat, and they no longer display the site selection or orientation behavior that are characteristic of smaller shrimp. The same individual thus

The kelp gunnel (*Apodichthys sanctaerosae*), an eel-like fish from Monterey Bay, California, inhabits beds of giant kelp (*Macrocystis* sp.) Photograph by Kawika Chetron. Used with permission, estate of Kawika Chetron, 2005.

switches between strategies depending on the limitations of its developmental stage and the features of its characteristic habitat.[7]

In object resemblance, Osbeck and Wallace recognized that the coloration of animals was a response to predation. No dietary or other immediate environmental influence could account for the elaborate structures that made fish and shrimp look like kelp or katydids look like leaves. Wallace and Darwin incorporated those observations into an account of biological change that has endured a century of scientific examination. Science advances with evidence from collecting and observing, but it solidifies with experimentation. Experiments like those of Hacker and Madin put historical observations to the test to determine the factors that influence the evolution of coloration.

3. Pepper and Smoke

MANY BIOLOGICAL CHANGES unfold before our eyes in real time. An octopus transforms the color and texture of its skin in a few seconds; a moth emerges from a cocoon over a couple of hours; a tadpole turns into a frog in a few weeks. But evolutionary change in most animals is a slower, more ponderous process. Small, chance events distributed over large populations unfold over generations, and the time span for change is determined primarily by the rate of reproduction. Only rarely does animal evolution occur so rapidly that it can be measured within a single human lifetime. The most celebrated incident was an instance of natural selection for concealing coloration, and it was first discovered, not by academic scientists, but by a legion of passionate amateurs.

Natural history was a popular sport in the late eighteenth century. Fashionable drawing rooms were filled with birds' nests, displays of exotic feathers, mounted animals, and specimens of shells, minerals, and flowers. Some of these curiosities were purchased, but many were the harvest of recreational expeditions to the countryside. Collecting insects, in particular, was a principal hobby of the era. The most dedicated collectors of moths and butterflies regularly met in the amateur entomological clubs of England, where associations of lepidopterists dated back to before 1700. Members exchanged specimens, shared information about collection and breeding techniques, and remarked on their acquisition of rare and unusual forms. Because they were alert to any variation in the appearance of their totem animals, it was amateur entomologists who first noticed the occurrence of novel, black forms of European moth species. The most thoroughly documented case was that of the peppered moth.[1]

Caterpillars of the peppered moth (*Biston betularia*) resemble twigs. Photograph by Stephen Dalton (Getty Images).

James Tutt illustrated the pale form of the adult peppered moth (*Biston betularia*) in 1896.

Peppered moths are camouflaged both as caterpillars and as adults. The caterpillars feed on various broad-leaved trees throughout Europe and North America. They strongly resemble brown or green twigs, matching the background color of the tree they are feeding on. For long periods during the day, they enhance their twig mimicry by adopting a rigid posture, holding themselves motionless at an angle away from a branch. Adult male peppered moths fly as far as six kilometers in a night in search of receptive females. During the day, the adults rest on the bark of tree trunks and larger limbs. The typical adult moth is pale, creamy white with a pattern of dark flecks and patches on its wings. On a tree trunk, it resembles a patch of pale lichen. In England, peppered moths are consumed by a variety of insect-eating birds, including great tits, hedge sparrows, European robins, spotted flycatchers, nuthatches, yellowhammers, and song thrushes.[2]

Before the Industrial Revolution, the pale form of the peppered moth was common throughout the range of the species, though there appear to have been small numbers of darker forms. After 1800, increased industrial development in the English midlands produced massive amounts of pollution from coal smoke. Lichens downwind from the factories were devastated by acid rain, and sticky soot blackened the bark of forest trees. Amateur entomologists first noted a novel, dark form of the peppered moth in these areas during the early 1800s. Its numbers rapidly expanded, eventually comprising most of the population in polluted forests near Manchester and Birmingham. In moth populations from unpolluted rural regions, however, the pale form remained common. When pollution controls were instituted during the 1970s, lichens began to recover, tree bark in industrial areas became lighter, and the abundance of the dark form began to decline. The dark moths are now uncommon throughout most of their range.[3]

In 1896, a respected English entomologist, James William Tutt, suggested that the occurrence of the darker *melanic* form of the peppered moth was the result of natural selection by predatory birds. He noted that in the areas where the black forms appeared, the trees were covered by dark sticky soot, and all of the light-colored lichens had been killed. A light-colored moth on a blackened tree would be "very conspicuous and would fall prey to the first bird that spied it out" (Tutt 1896, 305–6), while black moths would be more likely to escape. As a consequence, peppered moth populations in the polluted forests would gradually evolve to include more dark forms, while populations in less polluted, rural areas would remain pale.

Tutt's speculative explanation was only thirty years after the publication of the *Origin,* and the idea of natural selection by visual predators was far from universally accepted. There were plenty of other possible interpretations, including suggestions that the dark moths were more resistant to pollution, or that pollution acted directly on the moths, causing them to change color. Did birds actually eat enough moths to make a difference, and were they sufficiently discriminating to select between the dark and light forms? These and similar questions would require extensive experimental investigation, producing a small industry in peppered moth research over the following century.[4]

Some of the earliest specimens of the dark forms of the peppered moth (*Biston betularia*) are from the William Jones Collection, collected before 1818 and later donated to the Oxford University Museum of Natural History. Photograph by Alan Bond. Used with permission of the Hope Entomological Collections, Oxford University Museum of Natural History.

An enthusiastic young scientist named Bernard Kettlewell was offered a fellowship in 1948 at Oxford University to work with E. B. Ford, the director of the newly formed Department of Genetics. Ford studied genetic variation of natural populations, and he was particularly interested in finding someone to investigate peppered moths. He offered Kettlewell a part-time fellowship and a half-time position at Radcliffe Hospital, room in the Oxford museum, 400 pounds per year, and an invertebrate animal house for breeding work. For the next twenty years, Kettlewell would try to nail down whether the change in abundance of different color forms of the moths could be fully explained by bird predation.

A lifelong field lepidopterist, Kettlewell had studied medicine and zoology at Cambridge and briefly practiced medicine in Oxford before taking a position with a locust study project in South Africa. He was an ideal choice for the Oxford fellowship. Kettlewell was thoughtful, meticulous, and doggedly persistent, and he collaborated well with a wide range of other scientists. In addition, he was highly skilled at raising moths from their eggs in captivity, an essential requirement for experimenting on them in the field.[5]

By the time Kettlewell began working, it had been well established that the light and dark forms of peppered moths were, in fact, genetically distinct. Whatever else smoke pollution was doing, it was not directly producing dark coloration. Kettlewell's initial step, therefore, was to ask whether pollution and dark coloration were tightly interlinked over the full range of the species, or whether this just seemed to be true because of the dominance of dark-colored forms in some highly industrialized areas. For this, he made use of his connections in the amateur entomology community. Over nearly twenty years, Kettlewell gathered data from 170 observers scattered across England, producing a detailed map of the numbers of light and dark forms. The geographical variation and historical changes in these numbers corresponded well to the local intensities of industrial pollution, an observation that has been confirmed in numerous replications of his work. There was no way, however, that even a very thorough ecological survey could provide direct evidence of natural selection by birds. Kettlewell needed to run experiments.

The experiments relied on a crucial aspect of moth behavior. When live peppered moths are released onto a tree trunk in the morning, they quickly settle on a spot and then stick to it throughout the rest of the day. Kettlewell set out samples of live moths that tended either to match or contrast with the coloration of the surrounding bark, and then tested the effects of their background on the moths' subsequent survival. His first study asked whether dark moths were inherently harder to find in polluted forest. He released moths of both forms singly on trees in polluted and unpolluted forests, and then tested how well humans could locate them. Kettlewell concluded that dark moths were harder to detect on polluted trees, while pale moths were harder to find on unpolluted ones. He then examined whether birds would show the same effect.

In this second study, Kettlewell released pale and dark moths on the same and contrasting backgrounds, first in an aviary and later in polluted and unpolluted forests. He observed the action from a nearby blind. The predators' response was complicated: The data suggest that a large proportion of the kills were produced by only a few individual birds. But Kettlewell was able to show that "on the majority of occasions the more conspicuous of the two forms were all taken before any of the others" so he felt bird behavior was at least qualitatively similar to the judgments of human observers (Kettlewell 1956, 294).

In his third and most important experiment, Kettlewell addressed the question of whether mortality was actually higher among moths that did not match the background. He marked large numbers of male moths of both forms on the undersides of their bodies and then released them sparsely on trees in the early morning. For the next several nights, he ran traps using both lights and the scent of female moths to retrieve surviving individuals. The experiment was conducted first in an area near Birmingham where dark moths were common and the trees were covered with soot, and then in an unpolluted rural forest at Deanend in Dorset. Kettlewell reasoned that if significantly larger proportions of one of the two forms were recaptured, this would be evidence of differential mortality at that location. If more dark moths were recaptured in Birmingham while more light moths were recaptured in Dorset, the result would support the hypothesis that concealing coloration had survival value. This is exactly what he found. In both polluted and unpolluted woods, moths of the apparently more cryptic form were two to three times as likely to be recaptured. This indicated

Kettlewell released both light and dark forms of the peppered moth (*Biston betularia*) on trees. Both forms are shown on unpolluted bark. Photograph by Stephen Dalton (Getty Images).

Kettlewell's experiments on peppered moth survival in unpolluted woods were conducted at Deanend in Dorset in southern England. Photograph by Judy Diamond.

that dark moths survived longer on smoke-polluted trees and light moths survived longer on pristine, unpolluted ones.[6]

The geographical variation and historical changes in moth coloration in Kettlewell's survey correlated with the local intensities of industrial pollution, and the strength and direction of evolutionary change in his mark-and-recapture study have been confirmed in several replications of his experiments. There is no doubt that peppered moth coloration evolved in response to environmental change.

There were, however, lingering concerns about some details. It was pointed out, for example, that Kettlewell's moths were all released on tree trunks, even though most peppered moths appear to settle on the undersides of branches in the canopy. The moths were also released in higher numbers than would usually have been present in the forest, and they were not released in the proportions that were normally found in the area. Some scientists expressed concerns that Kettlewell set out his moths in daylight. Since peppered moths do not like to fly in the daytime, they must have landed as soon as possible after being released and stayed there all day. These quickly chosen locations might have been less concealing than ones that the moths would have chosen on their own if they had had the time. Finally, many of Kettlewell's moths had been raised in the laboratory, and it seemed possible that there were behavioral differences between lab-raised and wild moths in how well they could find places to hide.[7]

Many of these concerns were addressed in a careful replication of Kettlewell's research conducted by Michael Majerus, a professor of genetics at Cambridge University. Majerus collected peppered moths from woodlands to the west of Cambridge between 2001 and 2007, when the woods were still recovering from coal smoke pollution. During the six years of his study, he found that the frequency of the dark form in these woodlands declined from 12 percent to just over 1 percent. The dark moths were essentially disappearing altogether. The question Majerus asked was whether this decrease could be attributed to bird predation. He captured local peppered moths and raised others in the laboratory. All the moths were released in an experimental plot near Madingley Wood. Improving on Kettlewell's technique, Majerus released low densities of the moths, and the numbers of light and dark forms were close to their natural abundance. The moths were released on different parts of trees, in roughly the proportions that they used these locations in the wild. Although both lab-raised and wild-caught moths were used, their data were recorded and analyzed separately. Most importantly, the moths were released into large cages on the trees—one moth per cage—at sunset. The cages were removed forty minutes before sunrise the following morning, and each moth was watched for a subsequent four-hour period, recording all incidents of predation by birds.

Significantly more dark peppered moths than light ones were eaten by birds in the Cambridge forest. The birds' bias toward dark moths, moreover, was a very close match to the overall decrease in dark moths in the population. Majerus had carefully addressed each criticism of Kettlewell's design and had found that the change in numbers of light and dark moths in a single forest over six years could be fully accounted for by bird predation. Kettlewell's results had shown that the coloration of the peppered moth was the product of natural selection. Majerus had confirmed that the primary agency of selection was, in fact, visual predation by insectivorous birds, producing rapid evolution of color forms in synchrony with local environmental change. The narrative seems very clear and simple, much simpler than cases of color evolution where the process cannot be observed in real time. As Majerus remarked on his results, "The peppered moth story is easy to understand, because it involves things that we are familiar with: Vision and predation and birds and moths and pollution and camouflage and lunch and death" (Majerus 2007).

The rise and fall of the dark form of the peppered moth is still considered one of the best-established and clearest illustrations of adaptive evolution, and it provides a window into how evolution operates in nature. Sewell Wright, one of the founders of the field of population genetics, considered the example of industrial melanism "the clearest case in which a conspicuous evolutionary process has been actually observed" (Wright 1978, 186). Science often proceeds from the work of researchers like Michael Majerus, who painstakingly replicate earlier studies to verify the accuracy of the findings and thereby reveal how biological theories are tested, evaluated, and strengthened over time.[8]

4. Obscured by Patterns

ON A BITTERLY COLD evening in 1914, Pablo Picasso walked with Gertrude Stein along the Boulevard Raspail in Paris when a convoy of military vehicles drove past. World War I was just in its opening phases. These trucks and artillery were the first examples that Picasso had seen of the new military camouflage, in which various colors were combined in swirling patterns to obscure the outlines of the vehicle. Picasso excitedly proclaimed that the patterns on the trucks were an example of cubism, the avant-garde art movement that he and Stein had helped to create.[1]

Military equipment had traditionally been colored to provide a sense of tribal or national identity. The brilliant scarlet coats of the nineteenth-century British army, for example, were designed to be conspicuous, to intimidate the enemy by emphasizing the number and discipline of the force arrayed against them. But conspicuous coloration was a significant disadvantage in World War I, when the opposing sides were equipped with new technologies that could inflict horrific injuries on soldiers from great distances. Military strategists began to change the way they outfitted their troops and equipment, basing their new designs on the concealing color patterns of animals.[2]

The artist most responsible for the use of obscuring patterns in military camouflage was not Picasso, nor was he even remotely avant-garde. He was Abbott H. Thayer, a prominent American painter best known for his portrayals of ethereal women and children and his landscapes of the New Hampshire mountains. Thayer was also a master at producing detailed, natural renderings of native birds and mammals. He believed that only an artist could fully understand how animals use color patterns to become invisible, and he was convinced that camouflage was essential in modern warfare. During the war, Thayer carried on extensive correspondence with British and American government officials, sending them sketches of how ships should be painted to conceal them from the enemy and making suggestions for draping submarines in camouflage cloth. How much influence these arguments had is not clear, but both allied navies painted many of their warships in zebra stripes, in accordance with Thayer's principle of dazzle coloration. One of Thayer's last books was focused entirely on the application of concealing coloration to military camouflage.

Thayer's years of wildlife illustration had convinced him that concealment is often due as much to pattern—the arrangement of the splotches, speckles, stripes, and shading that make up an animal's visual appearance—as to the particular colors employed. His famous painting of a male peacock displayed against a blue sky and tropical foliage illustrated how a color pattern could conceal even apparently conspicuous animals in a sufficiently complex environment. Thayer contended that contrasting stripes that run out to the edge of an object fracture the visual outline, making it difficult to separate the object from its surroundings. He also thought that contrasting blobs of color within the outline of an object tend to adhere visually to matching parts of the background, thereby preventing the object from being recognized. Because both these patterns served to break up an image, they have come to be called disruptive coloration. Under the right circumstances, Thayer contended, this disruptive patterning can literally obliterate an animal's visual appearance.

Thayer was certain of these principles because he could demonstrate them in paintings. Art did not just represent reality for Thayer. Art was purer than reality, providing a direct visual

Thayer believed that all animals, given the right environmental conditions, were concealingly colored. His most famous painting is *Peacock in the Woods* (study for *Concealing-coloration in the Animal Kingdom*), which suggested that even an animal as conspicuous as a peacock (Indian peafowl, *Pavo cristatus*) could sometimes be hard to see. Used with permission, the Smithsonian American Art Museum.

understanding that could not be obtained from just looking at a real peacock strutting across a lawn. His approach was hardly scientific, but it was very effective: The principle of disruptive coloration became broadly accepted in the scientific literature.

Thayer's illustration, *Blue Jays in Winter* (study for *Concealing-coloration in the Animal Kingdom*). Used with permission, the Smithsonian American Art Museum.

Only recently have scientists conducted serious experimental tests of Thayer's principles, tests that rely not on artistic insight but on the perceptual abilities of real predators.[3]

Researchers in England have tested Thayer's hypotheses using carefully designed insect models placed in forests where they

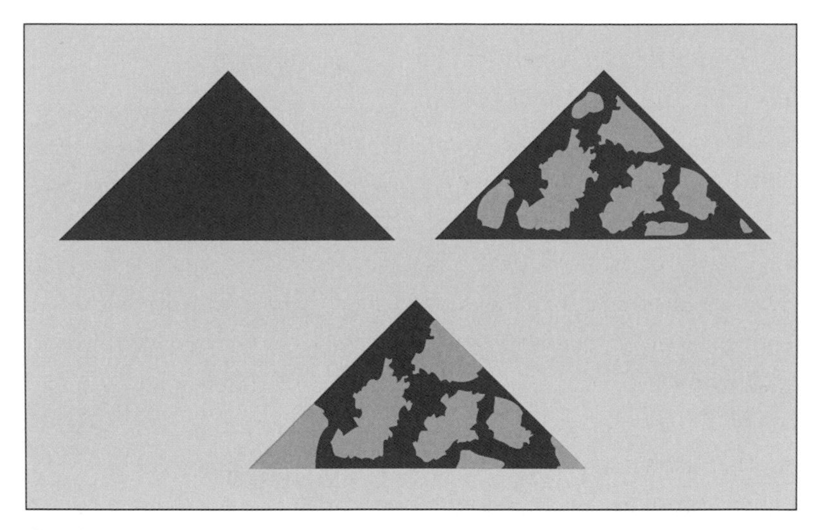

Martin Stevens created paper moth models to test the effectiveness of varying levels of disruptive patterns. The model in the upper left is uniformly colored in an average shade of tree bark. The one in the upper right has contrasting color patches that do not break up the edges. The one in the lower center, with patches running out to the edge, is disruptively patterned and harder for birds to find. Used with permission of Martin Stevens.

can be attacked by free-living predatory birds. Martin Stevens, at Cambridge University, and his colleague, Innes Cuthill, at the University of Bristol, wanted to determine exactly how disruptive color patterns serve to conceal prey. They created models of moths out of paper triangles, coloring them with barklike patterns. They then attached a tasty mealworm to each moth and pinned the models to trees in an English woodland. Over the next several days, they recorded which paper moths had lost their mealworms.

Stevens and Cuthill found that the models with patterns that disrupted their outline were the hardest for birds to detect and thus retained their mealworms much longer. The task was easier for birds when the pattern was shifted so it did not touch the edges. The solid-colored models were the easiest for birds to find. Disruptive coloration did not work well on its own, however. It was most effective when it made use of colors that were drawn from the palette of the surrounding bark. [4]

Thayer's most famous discovery was the visual pattern now called countershading, in which the upper surface of the animal is much darker than the lower one. This pattern is found in a wide range of animals from insects to fish, birds, mammals, and reptiles. Although dark backs and light bellies had been known from early times, classical writers assumed that it was due to some unknown effect of light or temperature on pigmentation. Thayer saw that the color gradation was the exact opposite of the shading that artists apply to two-dimensional images to make them appear three-dimensional and to give the illusion of depth to their paintings.

Put a uniformly colored wooden sphere in sunlight, and the top will be brightly illuminated, while the bottom is in shadow. This phenomenon is so consistent that the visual system immediately interprets the contrast as an indication of depth, causing three-dimensional objects to appear to stand out from the background. Thayer saw that the perceptual process could be made to work in reverse. If he painted a sphere with an inverted color gradient, making it darker on the top and lighter below, and then illuminated it from above, the natural illumination pattern and the painted gradient would cancel each other out, and the sphere would appear as a flat disk. If the sphere were painted in colors and patterns that match the background, countershading would serve to accentuate the resemblance, "and the spectator," Thayer wrote "seems to see right through the space really occupied by an opaque animal" (Thayer 1896a, 126).

As with his paintings of disruptively colored birds, Thayer felt that the road to undeniable truth lay in direct perceptual experience. He brought his "beautiful law of nature" before the scientific commu-

ABOVE: The lime hawk-moth (*Mimas tiliae*) illustrates a disruptive color pattern. Photograph by Garden Picture Library (Getty Images). **BELOW:** A countershaded willet (*Catoptrophus semipalmatus*) catches a snail. Photograph by Stan Osolinski (Getty Images).

nity by conducting demonstrations for scientific societies using three-dimensional sculptures. In 1896, on the lawn in front of what is now the Harvard Museum of Natural History, he demonstrated his principle of countershading to the American Ornithological Union. The secretary of the society described it in detail:

> Mr. Thayer placed three sweet potatoes, or objects of corresponding shape and size, horizontally on a wire a few inches above the ground. They were covered with some sticky material, and dry earth from the road on which they stood was sprinkled over them so that they would be the same color as the background. The two end ones were then painted white on the under side, and the white color was shaded up and gradually mixed with the brown of the sides. When viewed from a little distance these two end ones, which were white below, disappeared from sight, while the middle one stood out in strong relief and appeared much darker than it really was. (Sage 1897, 85)

Thayer had arrived at his principle in an artistic epiphany, free of any evolutionary inferences. His initial publications did not even mention natural selection. But he subsequently took his exhibits on a tour of natural history museums in England, where he met Alfred Russel Wallace and several of his colleagues. During that visit, Thayer became fully converted to an evolutionary perspective. He concluded that countershading and disruptive coloration should be recognized as "the most wonderful form of Darwin's great Law" (Thayer 1902, 597).[5]

Thayer had impressed scientists by demonstrating that countershading makes potatoes or wooden sculptures harder for humans to detect. But it was quite another task to make a case that countershading consistently reduces the detectability of concealingly colored prey. The range of animals that appear to be countershaded includes large marine predators, seabirds, snakes, and nocturnal mammals, species that would not obviously benefit from eliminating solar shadows. It would take experimental studies conducted under field conditions to convince modern scientists that countershading helped some animals to survive.

Observations of caterpillars by Leendert de Ruiter, a Dutch scientist, provided inferential support for Thayer's ideas. In the 1950s, de Ruiter carefully evaluated the countershading and behavior of seventeen concealingly colored species of caterpillars that feed on tree foliage in England. Roughly half of these species typically rest along the upper sides of branches and twigs, while the others usually hang upside-down from the lower sides. De Ruiter observed that the direction of countershading was perfectly correlated with the caterpillars' favored positions: Species that stayed on the branch tops were darker on their backs; those that hung underneath were darker on their bellies. And when he brought them into the lab and shone light on them from below, they reversed their preferred orientation. This suggested that countershaded caterpillars were choosing a position that minimized the effects of shadows. It was still necessary, however, to determine whether the countershading protected caterpillars from real predators.

A group of English biologists recently put Thayer's countershading theory to scientific test. They presented caterpillar models to birds in an undisturbed forest, asking whether the presence of countershading actually reduced predation. Hannah Rowland, from the University of Liverpool, constructed model caterpillars from cylinders of pastry dough. Each cylinder was colored either uniformly dark green, uniformly light green, dark above and light below (countershaded), or light above and dark below (reverse shaded). She pinned them along the upper surfaces of tree branches in an English forest, much as Stevens and Cuthill had done with their model moths. Over successive days, she returned to see which caterpillars had been eaten.

Rowland found that countershaded prey experienced much lower levels of predation than uniformly shaded or reverse-shaded prey. Countershaded prey were difficult to detect when their pattern could offset the effect of natural shadows, but not when they were presented upside-down. She repeated her experiments in a variety of environments and under different lighting conditions and consistently found that countershading reduced predation compared to uniformly colored, background-matching prey. Her results supported Thayer's hypothesis, but she cautiously noted that "the actual mechanisms by which countershading functions to reduce attacks by avian predators are still to be determined" (Rowland 2011, 67).[6]

Masters of Pattern

Concealment is often at least as much an effect of pattern as of color. Animal coloration usually appears as distinctive markings like the narrow stripes on grassland birds and insects, the spots on jungle cats, the leaf-sized patches of muted color on animals of the forest floor, or the fine mottled speckles on sandpipers and bottom-dwelling fishes. Such patterns echo the texture of the environment, matching the animal's surroundings in the size and shape of color patches, as well as in the particular hues and intensities of coloration. To envision how pattern contributes to concealment, think of a digital image of an animal against its background. The image is composed of pixels of specified colors, and adjacent pixels that are similar in color will appear to the eye as patches of varying size. How well the animal blends into its surroundings is a function of the match between the color of the patches in the background and those on the animal, but blending in is also affected by the relative sizes of the patches, regardless of their color. Animals process these two kinds of visual information, color and patch size, in separate parts of their nervous system, and the information is subsequently integrated to assess the degree of resemblance to the background.[7]

How animals integrate the information about color and patch size is most impressively illustrated by cuttlefish. Cuttlefish,

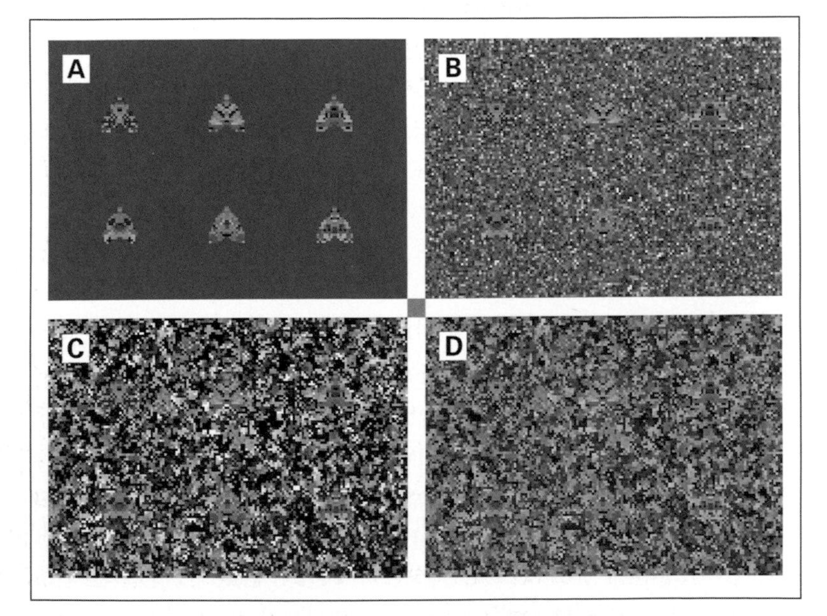

Can you find the moths? The six identical digital moths, shown in the same positions in each panel, are harder to see depending on how well their patterns and shades resemble their surroundings. In (A), they are shown on a uniform gray background. In (B), patches in the background are smaller than those in the moths but are the same shades of gray. In (C), the patches on the moths and the background are the same size, but their shades of gray do not match. In (D), both shades and patch sizes match. Image by Alan Bond.

which are free-swimming relatives of octopus and squid, are masters at generating patterns to match their environment. They have large, complex, sensitive eyes and a sophisticated control system that allows them to rapidly modify the patterns they display. When they detect changes in their surroundings, they can change the arrangement of light and dark patches on their body accordingly. At the Marine Biological Laboratory at Woods Hole in Massachusetts, Roger Hanlon studied how cuttlefish create their remarkable patterns. In one experiment, Hanlon and his colleagues placed cuttlefish in transparent aquaria on top of complex backgrounds, such as checkerboards, stripes, or spots. They photographed the cuttlefish to determine how well they matched the background patterns, and they timed how rapidly the animals could change from one pattern to another.

The change was immediate. Hanlon and his team found that cuttlefish use three major body pattern types for camouflage: uniform patterns on backgrounds with little or no contrast; mottled patterns on substrates with small, highly contrasting patches like small checkerboards; and disruptive patterns that break up their outline on substrates with high contrast and long, defined edges, like large checkerboards. In the lab, cuttlefish will camouflage on any substrate, natural or artificial, regardless of whether or not there is a predator in the tank.

Hanlon asked whether the cuttlefish had preferences among background types. To ensure that previous experience had no influence in their choice, the cuttlefish were raised in the lab on backgrounds of a solid color. First, the cuttlefish were placed on a substrate that was half white and half uniform gray. In this case, the animals clearly preferred the darker substrate. When they were presented with soft sand versus sand that had been glued to plastic so the animals could not burrow into it, the cuttlefish

Cuttlefish (*Sepia* sp.), shown here on coral from Indonesia, are superstars of rapid pattern matching. Photograph by Danita Delimont (Getty Images).

preferred the soft sand. When cuttlefish were given a choice of three different artificial backgrounds—uniform gray, small black-and-white checkerboard, and large checkerboard—they showed no preferences, and they still showed no preferences when given three natural backgrounds: sand, small shells, and large shells. Subsequent experiments showed that cuttlefish readily integrate multiple cues from a mixed substrate (like shells of different sizes) producing a mixed camouflage pattern.

Cuttlefish are unusual because they generate their own patterns even on backgrounds they have never encountered before, and they can achieve this with lightning speed. They change their color patterns with organs called chromatophores, bags of pigment with muscle fibers radiating out around them. Each chromatophore contains one type of pigment, either dark brown, orange, or yellow. Because the organ operates with muscular contractions, color change in cuttlefish happens as fast as you can snap your fingers. A similar system occurs in most of the cuttlefish's relatives, including octopuses and squids. Many fish have pattern-matching capabilities, particularly flounders and other flatfish that inhabit sandy ocean bottoms, but their chromatophores work on a different basis and are significantly slower. No other animal has the versatility or speed of pattern matching as the cuttlefish.[8]

Many Ways to Change Color

John Maynard Smith, a British mathematical biologist, outlined three evolutionary strategies to account for animal color patterns. The first strategy he termed genetically adapted coloration: When animals are adapted for life in particular conditions, and when the conditions seldom change within the lifetime of the animal, the coloration they display is fully specified in their genes. No matter what environment they are raised in, they develop the same appearance, and they do not change color when conditions vary. The sargassum animals with their elaborate seaweedlike structures and the light and dark forms of the peppered moths display genetically adapted coloration. Genetically adapted coloration is modified only very slowly, as a result of changes in gene frequencies in populations.

Maynard Smith termed the second strategy developmentally flexible coloration. When mammals, birds, and insects shed their outer layers of hair, feathers, or exoskeleton and replace them with new ones, they often shift their appearance. This is particularly common in response to seasonal changes. Arctic species, such as ptarmigans, arctic hares, arctic foxes, and stoats, alternate between white and brown coloration from one molt to another, in synchrony with seasonal changes in the background. Insects can switch their appearance between successive developmental stages based on cues received from the environment, such as light, temperature, or humidity. Caterpillars of the peppered moth, for example, adopt the appearance of their host plant based on the coloration of the reflected light. They molt their skins several times in the course of development, and each time, they have the opportunity to modify their appearance.

The third strategy Maynard Smith called physiologically versatile coloration. The instantaneous and varied color patterns of the cuttlefish, the disappearing acts of flatfish, and the background matching of some chameleons are all examples of animals that are able to modify their coloration almost continuously to match varying background conditions. In these species, color change is under the control of the nervous system, and the speed of response can be very fast. Many other fishes, amphibians, and crustaceans can also respond physiologically to changing environments, but the changes generally occur on a slower time scale because they are mediated by hormonal signals. African cichlid fishes, for example, change coloration as a function of the hormones that control their reproductive state. Males become conspicuously colored when they are holding territories, but when they are not breeding, they revert to a more cryptic coloration similar to females or juveniles.

The color strategy that an animal evolves depends the variability of its environment—how predictably the surroundings change and how rapidly the change occurs. Stable environments are associated with genetically adapted color patterns that remain constant throughout the life of the animal. Seasonally variable environments can favor color patterns that are triggered by either internal hormonal changes or external cues associated with the change in seasons. Highly variable, unpredictable environments represent a particular challenge for many animals. Some, like cuttlefish and chameleons, can instantly modify their coloration, but most animals do not have this option. The spectacular color-change capabilities of cuttlefish are an extension of the physiology they share with other cephalopod mollusks. Instant transformation is not possible for mice or birds, which lack the cuttlefish's genetic resources. As a result, coloration in many vertebrates that cope with extravagantly changeable environments is often a compromise among less satisfactory alternatives.[9]

Although color evolution is shaped by the environment, animals are not infinitely flexible. Each species is constrained to some degree by the genetic and developmental mechanisms that are part of its evolutionary legacy, and these limit the sorts of innovation that can evolve in response to environment change. The genetic blueprint within each species contains hidden wealth, a host of processes that can be repurposed to aid visual concealment. Animals may not achieve a perfect match to their environment, but through the course of their evolution, they often get surprisingly close.

Part Two

PERCEPTION

[T]here is always too great a tendency to endow animals with senses exactly similar to those possessed by ourselves.

FRANK E. BEDDARD, *Animal Coloration: An Account of the Principal Facts and Theories Relating to the Colours and Markings of Animals*

The kea (*Nestor notabilis*), a parrot from New Zealand's South Island, is a generalist predator that hunts insects and young birds. Keas scavenge meat from dead animals and forage on a variety of plant materials, including berries, roots, leaves, and nectar. The vision of keas, like all parrots, extends well into the ultraviolet. Parrots may be the only avian order in which ultraviolet-sensitive cones are found in all species (see Ödeen and Håstad 2003; Carvalho et al. 2011). Photograph by Judy Diamond.

5. Colors in the Mind

RICHARD FEYNMAN LOVED to teach through unconventional but accessible analogies. In one of his accounts of light and vision, the Nobel Prize–winning physicist proposed that we imagine a swimming pool on a hot summer day, with lots of people repeatedly jumping in, swimming about, and climbing out again. The waves produced by each swimmer spread out and bounce off the sides of the pool. They interact and intermingle, adding and subtracting to produce a choppy chaos all over the water surface. Suppose that there is an insect, perhaps something like a water strider, floating on the surface in a corner of the pool. Depending on where a disturbance occurs and how large it is, the waves that reach the water strider could be small ripples, with short distances between each successive wave crest, or large swashes, with long distances between peaks.

Feynman envisioned a very clever water strider that could read the patterns in the irregular, choppy water to figure out where and when people entered or left the pool and what they did while they were there. It seems incredible, but in fact, much of that information is implicitly present in the waves. Just as our eyes detect energy waves, the water strider detects water waves, and the task the water strider faces is not that different from what our eyes and our nervous system do when we look at something. Vision is the process of unpacking the energy waves that come to our eyes from all directions and using them to reconstruct what is happening around us.

Imagine that the water strider is only able to detect a limited range of wavelengths in the choppy chaos of the pool. Short ripples pass by without being detected. Waves that are too long aren't recognized; they just make the bug drift up and down like a cork in the ocean. We are in a similar position with respect to energy waves because our eyes can detect only a narrow band of wavelengths. The light that humans can see ranges from waves about 400 nanometers (nm) long, which we see as violet, to waves of about 700 nm, which we see as red. A nanometer is 1 billionth of a meter, so these wavelengths are less than the thickness of a single strand of silk from a spider's web.

We are surrounded by all sorts of energy waves, but their lengths are too short or too long for us to see. Shorter waves range from invisible ultraviolet down through X-rays to cosmic rays, which have wavelengths about the size of an atom. Slightly longer than visible light are the invisible infrared or heat waves. Longer still are microwaves, with lengths of about a centimeter, and radio and television signals, with lengths of about a meter. Some forms of energy produced in lightning strikes have wavelengths on the order of 100 thousand kilometers, about three times the circumference of the earth. Although visible light occupies only a tiny fraction of the spectrum of all possible wavelengths, there is quite a bit of it around. About half the energy we receive from the sun is in the visible portion of the range.

When he described energy waves, Feynman was talking about electromagnetic radiation, the term for the forms of energy that can be transmitted through space. Electromagnetic radiation consists of tiny packets called photons that have wavelike properties. There is no color in the physical world. What we call red is how our brains interpret the electrical signals our eyes transmit in response to photons of a particular wavelength. The color itself is all in the mind. Waves that are 700 nm long are not intrinsically "red," any more than are the signals from your favorite FM station or the X-rays that make an image of your molars.

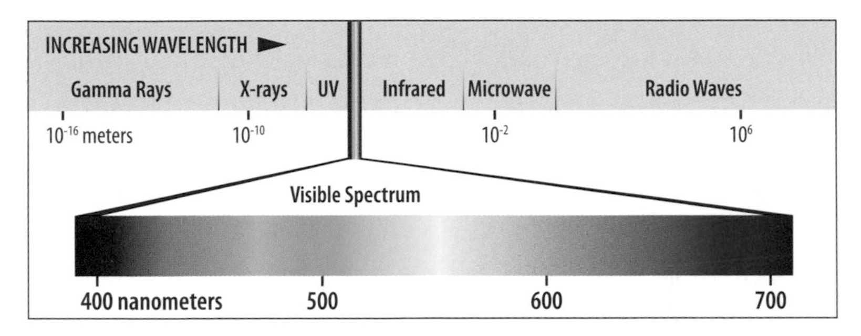

The part of the electromagnetic spectrum that is visible to humans is only a small portion of the entire range of energy. Illustration by Angie Fox.

What happens to an energy wave when it contacts an object depends on the material the object is made of and the length of the wave. Sometimes energy waves are transmitted through objects, like sunlight through air or radio waves through the walls of a house. Sometimes the waves are absorbed by an object, and the energy is turned into heat, like a black sweater warming in the sun. And sometimes the energy waves are reflected off objects, where they produce the perception of color.[1]

Animals have evolved a diverse set of chemicals and physical structures that produce the range of colors they display. Some of the colors of animals come from pigments, chemicals that selectively absorb some wavelengths of light and allow the others to be reflected. But color can also be produced in transparent, nonpigmented materials, as a result of the geometry and the physical properties of the material. Such structural coloration accounts for the rainbows in soap bubbles, the iridescence of opals, and many of the colors in the scales, feathers, and skin of animals.

The iridescent red feathers on the throats of hummingbirds, the luminous blue on the wings of the blue morpho, and the dramatic green of the peacock's tail are all produced by arrays of transparent structural molecules, particularly keratin and chitin. Keratin

The brilliant color in the wings of the blue morpho (*Morpho peleides*) results from interference of two thin layers of keratin separated by a tiny gap filled with air. Photograph by Raj Kamal (Getty Images).

is a fibrous protein found in skin, feathers, and scales and is the same material that makes up your fingernails. Chitin is the tough material in the exoskeleton of arthropods, such as beetles, cicadas, spiders, and crabs. When pits, blobs, and layers of chitin or keratin are arranged in regularly spaced arrays, they amplify the reflection of particular wavelengths, producing brilliant, spectrally pure colors.[2]

The visual appearance of most animals is not solely the result of structural coloration. It is also the product of biological pigments, a hodgepodge of otherwise unrelated compounds that are united only by a quirk in their molecular structure: They all have alternating saturated and unsaturated bonds between adjacent atoms. When the two types of bond repeatedly alternate down the molecule, they form a cloud of electrons that can readily store or release energy, even from low-intensity sources like visual light.

Chitin makes up the carapace and wings of the cicada (*Tibicen* sp.) shown above. Photograph by Judy Diamond.

The molecular structure of the pigment determines which wavelengths it will absorb and reflect. When pigments absorb all wavelengths equally, the result is a shade of white, gray, or black. Specific colors result from greater absorption in some regions of the spectrum. An animal that had no pigments at all would be nearly as transparent as water, with at most an opaque skeleton and silvery internal organs. There are a surprising number of animal species that come close to this level of transparency, eliminating all but a few essential biological pigments. One of the most unusual examples is the glass catfish, a predatory species from Southeast Asia. In a clean aquarium with good illumination, they look like living versions of plastic anatomical models, but in their native habitat of slow, peat-stained rivers, they are nearly invisible.[3]

The pigments that produce animal colors are often borrowed from biochemical pathways that have unrelated functions. In many cases, pigments serve a primary physiological purpose that only indirectly corresponds to their color-producing ability. This is particularly true of pigments called porphyrins, which help animals to convert and transport energy. Porphyrins consist of

The dark green facial color of the purple-crested turaco (*Tauraco porphyreolophus*) from southern Africa comes from turacoverdin, a porphyrin pigment. Photograph by Chris van Lennep (Getty Images).

The American flamingo (*Phoenicopterus ruber*) obtains its orange carotenoid pigments from the algae it eats. Photograph by Judy Diamond.

flat rings of four subunits, each of which is a ring of carbon and nitrogen atoms. This symmetrical multiringed structure gives the porphyrin molecule a property that is rare among biological chemicals—the ability to bind to metallic elements. The four nitrogen atoms pointed toward the center allow the porphyrin ring to firmly attach to metal ions, and the ions, in turn, enable the molecule to perform rapid transfers of electrons.

This is what makes porphyrins absolutely critical to life. A porphyrin containing magnesium, for example, is the business end of chlorophyll, the molecular complex in green plants responsible for photosynthesis, which converts light into chemical energy and releases oxygen. Porphyrins containing iron (in vertebrates) or copper (in many invertebrates) form the heme units that transport oxygen to tissues.

Porphyrins nonetheless make life colorful. Plants look green because their chlorophyll contains porphyrin rings that absorb energy from both the red and the ultraviolet, reflecting only wavelengths from the central blue-green portion of the spectrum. The hemoglobin in the blood of vertebrates contains a porphyrin that absorbs more from the ultraviolet than from the red end of the spectrum, and thus appears red. The blood of many invertebrates, such as lobsters and squids, looks blue green because their porphyrin absorbs both ultraviolet and red wavelengths and reflects blues and greens. In some species, such as earthworms, some marine snails, owls, and turacos, porphyrins play the additional role of providing the animal's color.[4]

The yellows in the feathers of canaries and in the stripes on monarch caterpillars, the oranges in the scales of goldfish and in the skin of red-spotted newts, or the reds of lady beetles and common starfish are all due to pigments called carotenoids. These chemicals consist of long chains of alternating saturated and unsaturated bonds between carbon atoms, often looped into six-carbon rings at each end. Like other pigments, the colors of carotenoids are secondary to other essential physiological functions. In plants, carotenoids are part of the mechanism of photosynthesis, absorbing the light wavelengths that chlorophyll does not require and dispersing their energy into the atmosphere as heat. Carotenoids absorb blue-green wavelengths and reflect reds and oranges, which is why plant carotenoids mostly look red to us (like the lycopene in tomatoes), yellow (like the xanthophylls in fall leaves), or orange (like the beta-carotene in carrots, sweet potatoes, and pumpkins).

Animals cannot synthesize carotenoids directly, but they transform carotenoid pigments obtained from plants in their diet. As with porphyrins, many of the essential functions of carotenoids in animals have little to do with external coloration. From dietary carotenoids, animals make vitamin A, which is used in growth, development, reproduction, and resistance to disease. But the most remarkable function of vitamin A is that it directly mediates visual perception. When a form of vitamin A, called retinal, is combined with the opsin protein, the resulting compound can translate light into electrical signals in the visual system. This carotenoid pigment enables all of us—jellyfish, octopuses, houseflies, guppies, and humans—to see. It is a marvelous symmetry that the same class of compounds serves both to generate brilliant colors in animals and to enable those colors to be seen.[5]

Colors that are used most often for concealment tend to be the greens of live foliage, the yellows and browns of dead leaves, or the umber of forest soil. Across different groups of animals, there is little consistency in how these colors are produced. Green spiders, for example, do not achieve greenness using the same biochemical mechanisms as green grasshoppers or green tree frogs. In many cases, the pigments used are waste products resulting from the breakdown of other pigments or nucleic acids. Instead of just being excreted, these compounds are modified and stored in specialized skin cells, where they contribute to the animal's external coloration. The visual appearance of a single color is often due to a combination of several mechanisms. Frogs, for example, have a blue-green structural layer that lies over a layer of yellow carotenoid. The reflected yellow light from the carotenoid is filtered through the translucent bluish skin, making the frog look green.[6]

The one major exception to this heterogeneous mixture of color mechanisms is melanin, which is far and away the most important pigment for producing concealing coloration. Melanin is a strange substance, a large molecule with a complicated network of interlinked units derived from tyrosine, an amino acid. Melanin absorbs most wavelengths of light, though it reflects a bit more in the red end of the spectrum. This ability to absorb light from across the spectrum underlies most of its biological functions.

Melanin is nearly ubiquitous and biologically ancient, recognizable even in the fossilized skin and feathers of dinosaurs. It occurs in all branches of the tree of life, although often in slightly different versions. Melanin is the ink that squids and octopuses release into the water when they are alarmed. It is also deposited in a layer that surrounds the light-sensitive cells in the eyes of vertebrates and insects, absorbing stray light rays and thereby improving visual sensitivity. And melanin in the skin absorbs destructive ultraviolet radiation, promotes immune function, and enables cold-blooded animals to increase their body temperature with heat from the sun.[7]

The two primary chemical forms of melanin are eumelanins, which are black to dark brown in color, and pheomelanins, which are golden yellow to reddish brown. The synthesis of both pigment

Hair color of all mammals, including the Virginia opossum (*Didelphis virginiana*) is the result of melanin. The red eyes are the result of light reflecting off blood vessels behind the retina. Photograph by Judy Diamond.

types begins with the same steps, but pheomelanins involve the addition of sulfur-containing compounds, and their subsequent synthetic pathways are different from those of eumelanins and are differently regulated. Mammalian hair includes no pigments other than melanins, so the possible range of colors should seem very familiar: blond, russet, dark brown, or black.

In vertebrates, melanin is synthesized in specialized cells called melanocytes. Under a microscope, melanocytes have a central blob and longish strands called dendrites dribbling out in all directions. Melanocytes are found only in skin but are closely related to nervous tissue, both in their origins and in their function. In these cells, melanin forms granules, like microscopic grains of black beach sand. In fish, amphibians, and reptiles, the melanocytes can aggregate the melanin granules in the center of the cell or spread them apart into the dendrites. When the granules are pulled together, more light passes through, so the skin becomes lighter in color. When they are spread out, melanin granules cover more area, and the skin becomes darker. The result is the ability to modify coloration in response to abrupt changes in the environment. Birds and mammals cannot change the distribution of their melanin as rapidly. They produce seasonal changes in the darkness of feathers or hair by transporting melanin granules into skin cells.[8]

Most animal colors result from pigments, structural colors, or a combination of the two. But color itself is not the last word when it comes to concealment. How well an animal matches its background is a function not only of color but of pattern. As long as an animal's coloration does not deviate too greatly from its surroundings, most of the fine-tuning of concealment is a matter of patterning. The evolution of concealing coloration can seem opportunistic, hijacking available chemical processes and structures and then repurposing them. But concealing coloration is also determined by the visual sensitivity and perceptual mechanisms of an animal's predators or its most important prey.[9]

6. The Beholder's Eye

THE EVIDENCE OF OUR own eyes has historically been the benchmark for interpreting animal coloration. But our visual system is one of many variations on the theme of seeing. Not all animals see the world the way we do. In fact, how an animal's colors appear to humans may be about the least important perspective, in the evolutionary scheme of things. Since coloration evolves in conformity with the perception of predators and prey, we should instead be asking how those animals see. Innes Cuthill put it succinctly, "When a bird and a human look at the sky, they really do *not* see the same color, even though the light reaching them is the same. This conclusion is not a philosophical point about the nature of subjective experience: it is a consequence of basic neurobiology" (Cuthill 2006, 6).

Of all the features of animals, eyes may be the most remarkable. The history of eyes goes back to the Cambrian Period, over five hundred million years ago. One predator called *Anomalocaris* had large compound eyes like those of modern insects and crustaceans, and it probably had more acute vision than most living invertebrates. *Anomalocaris* was not alone. Most Cambrian animals had clearly defined and well-developed eyes. Eyes are found in nearly all major groups of animals, having originated more or less independently about forty times over the course of evolution. They capture the information carried by light waves using every possible combination of physical mechanisms for focusing and projecting an image. These include multiple light sensitive devices with a single lens (the simple eyes of vertebrates and mollusks); single devices each with its own lens (the compound eyes of arthropods); and eyes that form images with pinholes (giant clams and chambered nautilus) and even parabolic mirrors (great scallops).[1]

All eyes contain photoreceptors, specialized nerve cells that convert light energy into electrical signals. The electrical signals travel to the brain, which reconstructs a representation of the visual field. A part of each photoreceptor is filled with a stack of folded layers of cell membrane, rather like a slice of baklava. Photons enter at one end and pass through the layers. Imbedded in the layered membranes is a compound made of the carotenoid pigment, retinal, bound to the opsin protein. Retinal has a long chain of alternating double and single bonds that allow it to absorb light energy. When it absorbs energy of a specific wavelength, it gets a kink in its chain of bonds, and because the opsin is attached to it, the shape of the opsin is also modified. So begins a series of falling dominos, where a photon produces a change in the retinal, which changes the opsin, which distorts the cell membrane. Finally, when enough opsins have been bent, the distorted membrane produces an electrical signal that travels to the brain to make an image. What is most amazing about this molecular mechanism is that all eyes work this way.

Opsins are ancient proteins that have been around since the beginnings of multicellular life, and retinal is the light-converting component in nearly all animals from jellyfish and flatworms to insects and humans. Across different groups of animals, there are structural variations in both molecules that have large effects on the particular wavelengths that produce molecular shape-shifting, but the basic mechanism is used everywhere. The other surprising constant is that the genes that control eye development, determining when and where a particular piece of skin and nervous tissue will transform into an eye, are very similar across a wide range of different organisms. These genes, called *PAX6*, are so similar that those from a mouse can trigger development of normal compound eyes in fruit flies. Other genes determine how the command to build an eye will play out in any given case, making single-lens eyes in vertebrates or compound eyes in flies. But high-level developmental control produces photoreceptors that always use the machinery of retinal and opsins. A lot of the basic molecular equipment for vision is part of the biological legacy of all animals.[2]

Within each major group of animals, the sensitivity of the eye and the range of wavelengths it responds to have been fine-tuned to the demands of a particular environment and way of life, resulting in eyes that are better for hunting mice in darkness or for locating ripe fruit in the treetops or for finding nectar in flowers. In vertebrates, the trade-off in visual systems is a compromise between sensitivity and acuity, between detecting fuzzy images in near darkness and seeing in high-resolution color in bright daylight. In the vertebrate eye, sensitivity and acuity are handled by separate, specialized photoreceptors called rods and cones. Rods detect the presence of light when the number of photons coming into the eye is limited. Cones provide sharp images like a high-resolution digital display, but they only work well when there is sufficient light.

The color of an object is a function of the set of wavelengths that are reflected back from the object to our eyes. Roses look red and leaves look green because they reflect wavelengths that our visual system encodes as red or green while absorbing or transmitting the rest of the light that falls on them. Cones in most animals come in multiple types that each respond to a range of different wavelengths. From the pattern of electrical responses of cones distributed across the retina, the visual system reconstructs an image in a full palette of colors.

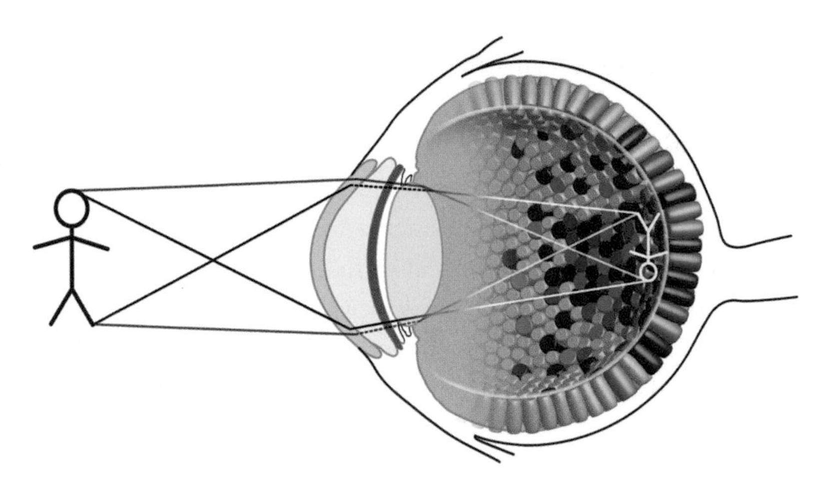

The human eye is a light imaging device with specialized receptor cells, called rods and cones. Rods respond to low light intensity, while cones respond to particular wavelengths of bright light. In this drawing, rods are shown in gray; cones are in blue, green and red. Illustration by Angie Fox.

Cones do not provide a graded signal: Each cell is just either on or off, like a light switch. So an object that reflects a balanced mixture of 620 nm (red) and 540 nm (green) light waves will turn on both medium- and long-wave cones and will appear to human observers to be yellow, indistinguishable from another object that actually reflects only 580 nm light (which also turns on both cone types). This will be true even if the red-green object absorbs all the actual 580 nm light that falls on it, so that no light from the yellow part of the spectrum is reflected at all. This is the principle used in color monitors and digital cameras, which emulate a range of natural colors by mixing graded intensities of just three specific wavelengths of red, green, and blue light. A picture of a banana on a computer screen may look yellow to our eyes, but its spectrum will be totally different from that of a real banana. Rods have only a single pigment type, like a black-and-white television, and from these the visual system can only interpret shades of gray. Vision in low-light conditions, when only rods can operate, does not provide color information, which is why you see colors poorly at night.

Old-world primates, such as humans, have three types of cones. S-cones have a peak light absorption in the short wavelengths around 420 nm, which appears dark blue; M-cones detect medium wavelengths around 530 nm, which looks green; and L-cones detect longer wavelengths around 560 nm, which appears red. This three-cone, or trichromatic, visual system allows humans and other apes to see much of the diversity of colors

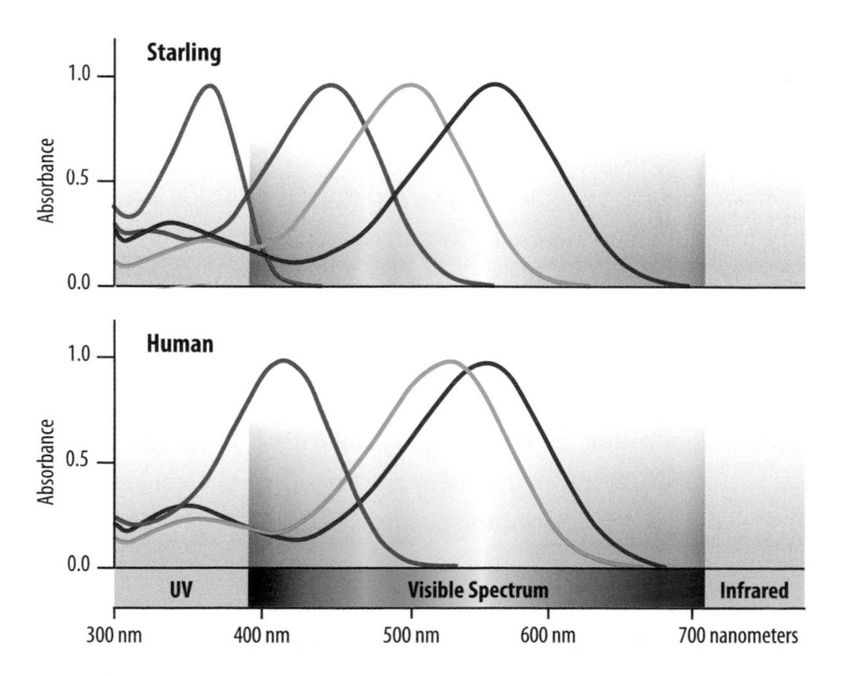

There are four different types of cones in the eye of the starling (*Sturnus vulgaris*), while human eyes have only three types. Each cone type is sensitive to a different range of wavelengths. The four pigments in the starling's cones extend their range of color vision into the ultraviolet, which most humans cannot see. Starlings detect peak wavelengths at around 360 nm (UV), 450 nm (blue), 500 nm (green), and 560 nm (red), and they have a double cone that overlaps with the green and red. In comparison, human cones detect peak wavelengths at around 420 nm (blue), 530 nm (green), and 560 nm (red). Illustration by Angie Fox, redrawn from Cuthill (2006).

contained in natural light. Honeybees, like most other insects, are also trichromats, but the wavelengths they see best are shifted to the ultraviolet. Honeybee eye pigments detect ultraviolet light in addition to blue and green. This gives them fine color discrimination at short wavelengths but with the trade-off that they do not see long wavelengths well. Many red flowers look black to them.[3]

What works for apes and some insects is not sufficient for many birds, reptiles, and freshwater fish. These animals are tetrachromats, with four different cone types. The cones of starlings, for example, respond to short (blue), medium (green), and long (red) wavelengths, plus a cone that runs well into the ultraviolet and a double cone that overlaps both long and medium single cones. This strategy is taken to extremes in some invertebrates: Asian swallowtail butterflies have eight receptor types, and mantis shrimp have ten, along with an array of different color filters. The eyes of these animals have made a trade-off: By having more types of color pigments with better color vision, they see less well under reduced light. Vision that involves three or more cone types is found mainly in animals that are active during the day and live in brightly lit environments.

Full color vision is not a high priority for most mammals. Except for primates, mammals are generally dichromats, with only two cone types. Ground squirrels, dogs, spotted hyenas, and many other species have eyes with relatively poor color discrimination, but they see well in the reduced light of dawn and dusk. Domestic cattle, for example, have two cone types: one detects wavelengths around 450 nm in the blue and the other peaks around 555 nm in the green. Although cows can discriminate colors, they can't do so with precision. Like color-blind humans, most mammals can't distinguish well among green, yellow, orange, and red.

The visual systems of animals range from monochromatic to polychromatic. **CLOCKWISE FROM TOP LEFT:** The southern elephant seal (*Mirounga leonina*) is a monchromat. Photograph by Seth Resnick (Getty Images). The parrotfish (*Scarus* sp.) is a dichromat. Photograph by Chris Newbert (Getty Images). The robber fly (*Holcocephala fusca*) is a trichromat. Photograph by Gustavo Mazzarollo (Getty Images). The chimpanzee (*Pan troglodytes*) is also a trichromat. Photograph by Frank Schneidermeyer (Getty Images). The blue jay (*Cyanicitta cristata*) is a tetrachromat. Photograph by Judy Diamond. The mantis shrimp (*Odontodactylus scyllarus*) is a polychromat with one of the most complex arrays of visual pigments of any animal. Photograph by Chris Newbert (Getty Images).

Dichromatic visual systems are most common in groups that evolved in environments with low light intensity and limited spectral range. Color resolution is sacrificed in these species for greater sensitivity by increasing the density of rod receptors. This has occurred not only in many mammals, but also in snakes and marine fishes. There are some animals, such as owls, seals, whales, and hamsters, that are monochromats, with only a single cone pigment, meaning that they cannot discriminate colors at all. All of these have rod-dominant retinas and forage in dark environments.

Animals inherit the features of their eyes as part of the shared traits that characterize their species, traits that are fine-tuned to the requirements of the environment in which they live. For example, fish from shallow, clear waters live in something close to natural sunshine. They are generally tetrachromatic with large numbers of cones, giving them high acuity vision with good color discrimination even at long wavelengths. Fish from moderate ocean depths live out their lives in a blue-green world from which most of the longer wavelengths have been eliminated. They tend to have dichromatic, blue-shifted retinas and a much greater density of rods. Fish from the abyssal regions of the ocean live forever in the dark. Some have pure rod retinas, because for them sensitivity is the only real concern.[4]

The visual system of a predator has consequences for the effectiveness of concealing coloration. Our elkhound loved to chase the cottontail rabbits that occasionally found their way into our backyard. The rabbits figured this out in short order. So as soon as the dog came out onto the porch, a rabbit that had been eagerly munching on clover would freeze, crouching close to the ground and flattening its ears against its back. The rabbit remained perfectly obvious to a human observer—a brown animal with

The spotted hyena (*Crocuta crocuta*) shown here in the Masai Mara National Reserve in Kenya, has dichromatic vision with only two kinds of photoreceptors. Its color vision may not be as acute as that of birds, but this predator sees well in low light. Photograph by Judy Diamond.

large, dark eyes does not disappear on a bright green lawn. But the dog was invariably baffled. Even when she was encouraged to search for the rabbit, she would just stare out at the lawn and jump around excitedly. When the rabbit finally became sufficiently alarmed and took off running, the dog would be after it like a shot. As long as the prey was motionless, the dog simply couldn't see it. To a dichromatic color-blind mammal with a rod-rich retina, a reddish-brown rabbit in full daylight is not easily distinguished from grass. In the seesaw of evolutionary trade-offs, the predatory canine has sacrificed color for being able to see in the dark, and it relies instead on its ability to detect movement. The rabbit's responses take advantage of the predator's poor color vision, and the ability to remain motionless becomes an enduring survival strategy.[5]

Predators can exert a powerful influence on the evolution of concealing coloration of their prey. But sometimes it is the prey that are the primary drivers of the color evolution of their predators. Take, for example, the crab spiders that lurk among the petals of showy flowers. The spiders are vigilant predators that ambush nectar-feeding insects, particularly bees and flies. They sit motionless, waiting for their prey with their first pairs of legs spread wide like the jaws of a leghold trap. When an insect lands on the flower, the spider closes in rapidly and delivers a killing bite. Their poison is fast acting: Crab spiders have been known to kill and consume bald-faced hornets, which are several times their size and are, themselves, formidable predators.

Sit-and-wait predators are frequently concealingly colored, but crab spiders have an additional trick in their predatory repertoire. Over a period of days, female crab spiders gradually switch their coloration from white to yellow, or vice versa, in response to the color of the blossom they inhabit. The overall reflectance spectrum of the spider pigment generally provides a good match to white or yellow flowers. There are striking anomalies, however. *Misumena* spiders are often seen on roses, where they gradually turn white. This seems contradictory for an animal that is supposed to be concealingly colored, but spiders, like bees, cannot see red. The spider's color shift on rose petals is apparently stimu-

ABOVE: The goldenrod crab spider (*Misumena vatia*) changes its color to match a yellow flower. Photograph by Nature's Inc (Getty Images). **BELOW:** The spider turns white on a rose because it cannot see red. Photograph by Michael Quinton (Getty Images).

lated by the ultraviolet component of the rose's reflectance, which humans are unable to detect.

Birds, with their tetrachromatic vision, can see the full range of wavelengths from ultraviolet to red, so most of them can see the spiders regardless of what flowers they are on. But spider researchers have looked carefully for evidence that *Misumena* is preyed upon by birds and have found very little sign of it. They have concluded that the color of the spiders must be mainly adapted to the vision of their insect prey, particularly bees, wasps, and flies. Several studies have found evidence that bees will notice the presence of crab spiders when they are not concealingly colored and will avoid those flowers, or they learn to avoid flowers where they had previously had bad spider experiences. The color match is not always perfect, but it appears to be good enough to fool some bees some of the time. Flies, which are less discerning than bees, can be fooled pretty much all of the time. Concealing coloration in crab spiders is, therefore, very much in the eye of the beholder, and the relevant beholder in this case is not one of the spider's predators but its most common prey.

Crab spiders have also developed behavioral strategies to maximize the likelihood of capturing their prey. They constantly switch from one flower to another, moving on if too few insects show up, and they are relatively open minded about the colors of flowers they choose. There is no clear evidence that they spontaneously pick yellow or white flowers over pink or blue ones. In fact, spiders are perfectly willing to sit on anything that brings in lunch. So the physiology of color change is just one part of an ecological interaction that includes the perceptual abilities of the local insects, the ability of bees to learn to avoid spider-occupied flowers, and the ability of spiders to judge a rewarding hunting location. No one should ever assume that a spider's life is simple.[6]

Because the human visual system is not well suited to evaluating animal coloration, we are left with a methodological problem: How does one determine quantitatively the appearance of one species when seen through the visual system of another? John Endler is an evolutionary biologist with a knack for designing useful analytical tools. He developed a method for describing animal colors that is grounded in the vision of the predators and prey themselves. Rather than depending on the limitations of human vision, Endler measures the effects of a predator's visual system on the detection of a prey animal, given the background color reflectance and the nature of the ambient light. Using this analytical approach, scientists can calculate the degree of color contrast between, for example, a mouse and its background of leaf litter under the actual conditions in which it might be seen by a hawk flying past at dusk. The system includes the effects of the absorption spectra of the predator's photoreceptors, so it considers not just the average difference between the mouse and the leaf litter under particular light conditions, but also the difference as it would be perceived by the visual system of the hawk.

In the evolutionary contest of predators and prey, poorer hunters and incompetent fugitives are weeded out. But ecology does not stage a dance with only two partners. Predators tend to be opportunistic in their choice of prey and will attack anything that could potentially be made into a meal. So, over the course of their evolution, most prey species are likely to have been influenced by the perceptual abilities of more than just a single species of predator. A deer mouse living in an overgrown meadow might be hunted in the daytime by tetrachromatic red-tailed hawks and trichromatic bull snakes. During the night, it would have to evade dichromatic red foxes and monochromatic great horned owls.

Predicting the outcome of selection in such multipredator

and prey ecologies seems like an impossible task, but things may be simpler than they appear. Daniel Osorio, a biologist at the University of Sussex, has carefully studied the visual systems of a number of species, from chickens to cuttlefish. He concludes that most animals rely on the perception of brightness or intensity as much as that of color. Intensity is actually more informative than color about changes along the borders of objects or where one pattern switches over to another one. As long as the mouse's coloration does not deviate too greatly from that of the background, its predators will rely mainly on the brightness of its coloration and the texturing of its fur to distinguish the mouse from its surroundings and to recognize it as prey.[7]

How an animal's coloration is produced and how its ecological partners perceive those colors are only the beginning of the story. A series of remarkable discoveries in desert habitats have helped to clarify the dynamics of color evolution. The conceptual advance that enabled these discoveries was born in the interaction of two fields of biology that were unimaginable to Wallace and Darwin—molecular biology and population genetics.

Part Three

ISOLATION

As touching Serpents, wee see it ordinarie that for the most part they are of the colour of the earth wherein they lie hidden: and an infinite number of sorts there be of them.

Gaius Plinius Secundus, *The Historie of the World*

Can you find the desert mantid (*Eremiaphila* sp.) photographed here in Jordan? Photograph used with permission, Dr. Matthias Borer, Curator, Department of Entomology, Natural History Museum, Neuchâtel, Switzerland.

7. Desert Islands

NAPOLEON BONAPARTE INVADED Egypt in 1798 with twenty-seven warships, one hundred cannons and siege guns, and thirty thousand seasoned French soldiers. But the young general was not simply out for conquest and glory. He expected a cultural exchange, bringing to Egypt the benefits of French scientific progress and returning to France with a narrative of the land, art, history, and customs of the ancient Middle East. So Napoleon brought with him a large group of scientists, artists, and engineers that constituted his *corps de savants*. Their task was to learn everything they could about Egyptian civilization, to teach the Egyptians about modern agriculture and engineering, and to survey the animal and plant life of the region.

A twenty-one-year-old naturalist, Jules César de Savigny, was appointed to catalog the Egyptian insects and other invertebrates. Following along with the invading army, Savigny amassed a huge collection of specimens and numerous detailed drawings. In some of the driest and most barren desert areas, often many days' travel from the Nile, the young explorer found strange, short-bodied praying mantids. In Savigny's engravings, desert mantids look like stubby-bodied, long-legged crickets. Only their huge eyes and their spiny grasping forelegs suggested their relationship to the more familiar mantids of Europe. The insects behaved much like tiger beetles, pausing for long periods and then abruptly racing across the desert pavement in pursuit of prey.[1]

The studies of the savants were a success, but the invasion was a debacle. Napoleon's fleet was destroyed by the English navy, and his army, ravaged by dehydration, plague, and guerrilla warfare, straggled back to France. The corps de savants, in contrast, came home with extensive notes and artwork. The French government ultimately published this material beginning in 1809 in a monumental encyclopedia entitled *La Description de l'Égypte,* and among the insects, the desert mantids received star billing. Later explorers were amazed to find that each of these insects precisely matched the coloration of its local soil. In one rocky wash, the mantids were the same dusty reddish brown as the surrounding gravel and clay. Scarcely a hundred paces away, on a salt flat encrusted with brilliantly white calcium carbonate, the mantids were pure white.[2]

The evidence of Savigny's mantids was a compelling demonstration of natural selection. Unrelated animals, like Tristram's birds and lizards, could have come to display a common coloration for a variety of nonevolutionary reasons, ranging from environmental effects to pure coincidence. But the mantids were all the same species and were all living under the same climatic conditions, and despite their similarities, neighboring mantid populations had diverged, taking on the appearance of distinctively colored patches of soil. Nonevolutionary explanations could not readily account for such divergent coloration. The idea of geographic isolation, where populations of a species occupy islands of visually distinctive habitat in otherwise uniform environments, would ultimately help to explain why divergent color resemblance has evolved in some cases and not in others. Significant progress has been made toward understanding the evolution of animal coloration, and some of the most important steps have resulted from research in desert environments.[3]

Figure 7.1 Desert mantids (*Eremiaphila* sp.) are widely distributed throughout Northern Africa and the Middle East. **ABOVE:** A mantid in Iran. Photograph by Mark Moffett (Getty Images). **BELOW:** A mantid in Jordan. Used with permission, Dr. Matthias Borer, Curator, Department of Entomology, Natural History Museum, Neuchâtel, Switzerland

A Mountain in the Desert

A party of scientists set up a field camp in the summer of 1889 at the base of the San Francisco Peaks, the eroded remnants of a million-year-old volcano in northwestern Arizona. The expedition's leader was C. Hart Merriam, physician and chief scientist for the Division of Biological Survey at the U.S. Department of Agriculture. The scientists were charged with assembling a comprehensive catalog of North American vertebrates. To accomplish the task, Merriam received only $600 from the Department of Agriculture to fund his entire expedition, which had to cover the cost of train travel, tents, and supplies, of renting horses and burros, and of hiring one cook and one wrangler for the three months he would spend in the field. Eventually, the party would survey the entire Colorado Plateau north to the Grand Canyon and east through the Painted Desert, an area of 30,000 square kilometers spanning over 2,300 meters in altitude.

Merriam's primary interest was in delineating life zones, which he thought of as coherent ecological communities that succeed one another along an altitudinal gradient. Like Mt. Rainier in Washington or Mt. Fuji in Japan, the San Francisco Peaks were originally a stratovolcano, a conical mountain with steep slopes formed from many successive layers of lava, ash, and mud flows over hundreds of thousands of years. Even after the erosion of the primary cone, the peaks still provide a spectacular gradient in habitats, from alpine tundra above treeline through sparse, high-altitude woodlands of Rocky Mountain bristlecone pine and Engelmann spruce to dark forests of Douglas fir and ponderosa pine on the lower slopes. Below the old volcano, pinyon pines and Utah juniper grow in dark umber soils across an outflow plateau that gradually declines in elevation. To the northeast, the outflow

Merriam described ecological life zones that changed with altitude on Humphrey's Peak, the largest of the San Francisco Peaks and the highest natural feature in Arizona. Photograph by Fotosearch Value (Getty Images).

terminates in the pale ocher and terracotta sands of the Little Colorado River Basin. Here, Merriam found a diverse Sonoran Desert community centered around fourwing saltbush, spineless horsebrush, narrowleaf yucca, and Mormon tea.

Merriam was a taxonomic splitter, inclined to draw fine distinctions between related groups of animals and to enshrine them in the literature as separate species or varieties. The most notorious example was his treatment of the brown and grizzly bears of North America, which he divided into eighty-four named forms based on skulls and skins from museum collections. Many of his diagnostic features later proved to be differences in the sex or age of the individuals, and the thicket of Merriam's bear nomenclature has since been pruned down to a single, morphologically variable species, *Ursus arctos,* distributed in five distinctive forms. Although his taxonomy has long gone out of style,

Merriam's approach was helpful in improving the understanding of animal coloration. Because he was alert to the possibility of subtle differences between related animals from different life zones, he paid careful attention to coloration patterns and how they corresponded to the colors of the surrounding soil and rocks. Using guns and his newly invented Cyclone mousetrap, Merriam collected numerous rodents from all the communities he surveyed.[4]

In several cases, Merriam found clear differences in coloration between related animals that lived on the dark lava soil under the pinyon pines and those that inhabited the nearby pale desert lowlands. For example, Merriam described the spotted ground squirrels from the lava soil as being a "dark russet hazel" with clearly defined spots, while the same species from the desert plains was a spotless, buffy clay color. The same contrast was seen in grasshopper mice. Their color on the lava outflow was dark, almost black, while the variety from the desert lowlands was pale, tawny cinnamon. Plains pocket mice from the lava soil were a uniform sooty brown on their backs, and Merriam predicted that a corresponding, pale ocher form would be found on the desert plains (and it was, in 1932).

The match of coat color to habitat was apparent among not only the small rodents but also the lizards. In the cedar and pine belts of the San Francisco Peaks, short-horned lizards matched the uniform dark brown color of the soil, but near the rim of the Grand Canyon, where the ground is covered with white, red, and brown sandstone pebbles, the lizards matched those surroundings so perfectly they could only be detected when they moved.[5]

Merriam was a steadfast Darwinist, acutely aware of the evolutionary implications of his observations. He noted that rodents in adjoining habitats, although they showed extremes of color

Short-horned lizards *(Phrynosoma hernandesi)* vary widely in color and pattern, corresponding to their local backgrounds. They are shown on gravel (above) and in rocky habitat (below). Photographs by Jeff Judd.

variation, had clearly descended from a common ancestor. Like Tristram before him, Merriam believed that predation was a major selective force in determining the color of desert animals. But Merriam's good friend, Theodore Roosevelt, was unconvinced. Roosevelt was less impressed with matching coloration than with the great variation in color among different species in the same habitat. Some of the animals on the San Francisco Peaks seemed to show adaptation to their local soils, but others were of much the same appearance wherever they occurred. Just out of the office of the presidency, Roosevelt wrote in a 1911 letter to Charles Kofoid, chair of zoology at the University of California at Berkeley: "Where a number of different species utterly differently colored exist with equal success, two things are sure; first, that if one of them is protectively colored, the others are not; and second, that this protective coloration must be of very small consequence compared with other features in enabling the animal to thrive" (Roosevelt 1911a, 96). Merriam's life zones demonstrated that coloration in some species could be tied to local differences in habitat. But it would be another American desert that would ultimately compel an evolutionary account of the color differences between populations and clarify the underlying selective mechanism.[6]

Land of Contrast

The Chihuahuan Desert consists of seemingly endless vistas of creosote bush, lechuguilla agave, and bunch grasses. But at its northern limit, an unusual combination of topography, geology, and climate has produced the largest expanse of gypsum dunes in the world. Gypsum is a mineral form of calcium sulfate, the same soft, white material that is used to make sheets of drywall and blackboard chalk. This region of dunes, called the Tularosa Basin, was produced by uplift of the surrounding mountain ranges, so there is no outlet for surface water. When late summer thunderstorms fall on the western mountains, the gypsum dissolves in the runoff and concentrates in a playa, an ephemeral lake at the southern end of the basin. As the water evaporates, the mineral recrystallizes, and the crystals are blown by the winds and piled up into dunes.

The Tularosa has been used for weapons testing since 1945, when the first atomic bomb was detonated there, but about half of the 700-square-kilometer dune field is administered by the National Park Service as White Sands National Monument. Although gypsum dunes are dry, unstable, and almost free of nutrients, a diverse community thrives in the white sands. Most of the plant cover is concentrated in the flat bottoms between the dunes, but several woody shrubs, particularly desert sumac and soaptree yucca, can survive on the dune crests, sending roots

At White Sands National Monument, New Mexico, plants are abundant in the flat areas between the dunes, where their roots can more easily reach water. Some populations of animals that live in the dunes have evolved distinctly pale coloration. Photograph by Judy Diamond.

down many meters into the soil below. As on the nearby desert plains, the plants provide habitat for rodents, insects, and lizards, but the gypsum dunes are distinct from the surrounding area in one important respect: Gypsum sand is white, unbelievably, astonishingly white, almost indistinguishable from refined sugar or fresh snow. Because of the size of the dune fields and the brilliance of their reflected light, White Sands National Monument is clearly visible from earth orbit.

Animals from the surrounding desert that are characteristically light brown, yellow, or gray occur as pale variants on the gypsum dunes. Throughout much of the Great Plains, for example, plains pocket mice are common in sandy areas, where they typically have yellowish-brown silky fur. The pocket mice that live in the gypsum sands, however, are nearly as white as albino lab mice. Southern plains woodrats, another sand-loving species, are dark bluish gray throughout most of their range in west Texas and eastern New Mexico, but the white sands variety is a pale, slightly smoky color. The three most abundant lizards in the gypsum dunes—the lesser earless lizard, the eastern fence lizard, and the little striped whiptail—are pale ash gray to white with minimal distinguishing patterning. Outside the dunes, they are dark gray and mottled or longitudinally striped with black. Ghostly Couch's spadefoot toads, with only their large, dark eyes distinguishable from the surrounding sand, emerge around the base of the dunes in large numbers after the first rains of the season. White Sands camel crickets wander the dunes at night and feed on dried vegetation. They are almost entirely without pigment, making their internal organs plainly visible through their translucent cuticle. Even just three kilometers outside the dune area, however, these crickets are fully pigmented, light brown with darker bands.[7]

Pale forms are confined to the gypsum dunes of the Tularosa Basin. **ABOVE**: The lesser earless lizard (*Holbrookia maculata*). Photograph by Judy Diamond. **BELOW**: The White Sands camel cricket (*Ammobaenites arenicolus*). Photograph by David Bustos, National Park Service, White Sands National Monument.

This is an impressive array of whitened animals, but what makes the Tularosa truly remarkable is that it includes both stark white and pitch black, both the yin and the yang of soil coloration. Just over 20 kilometers to the north of the dunes is the Carrizozo Malpais, a basaltic lava flow that covers over 300 square kilometers of desert plains to a depth of 10 to 15 meters. The lava is rough, with sharp ridges, ropelike flow areas, lava tubes, and large, broken bubbles, all crisscrossed with deep cracks and fissures. It is the result of several eruptions of Little Black Peak at the north end of the Tularosa from about five thousand years ago, making it the youngest lava field in the continental United States. Its surface has not yet been worn down by erosion, and as a result, the lava flow is as dark as asphalt. It appears in satellite images as a huge black pool, strongly contrasting with the pallid terracotta of the immediately surrounding desert. Despite its unpromising appearance, the Carrizozo Malpais is far from sterile. Soil has blown into crevices and pockets in the lava, supporting a diverse plant community of creosote bush, honey mesquite, yellow agave, fourwing saltbush, and several species of cactus and providing habitat for a number of different rodents, lizards, and arthropods.

In keeping with its asphalt hue, the Carrizozo sustains many species that are much darker than their relatives in the surrounding desert. Rock pocket mice are found on the lava flow, as well as on other rock outcrops scattered throughout the Southwest. They are highly variable in coloration across their range, generally tending toward yellow or reddish brown, but on the Carrizozo lava, their fur is uniformly coal black. White-throated woodrats, a rock-loving species closely related to the woodrats of the white sands, are brown on nearby outcrops; in the Carrizozo, they are completely black, except for a small patch of white on their

ABOVE: The black lava at the Valley of Fires in the Carrizozo Malpais in New Mexico forms a stark contrast to nearby White Sands National Monument. Photograph by Alan Bond. **BELOW:** The eastern fence lizard (*Sceloporus undulatus*) on the Carrizozo lava is darker than individuals of the same species found in surrounding areas. Photograph by Judy Diamond.

throats. The local rock squirrels and northern rock mice occur in similarly darkened forms, as do tree lizards, side-blotched lizards, and collared lizards. The most impressive color variation by far, however, is shown by the eastern fence lizards, which are common and broadly distributed throughout the Tularosa Basin. In the surrounding desert, these reptiles are grayish brown with lighter longitudinal stripes. In the white sands, a pale variety can be found, while on the lava flow, the lizards are dark charcoal gray.[8]

Naturalists from museums in Michigan and Berkeley in the 1930s conducted thorough field surveys in the Tularosa and neighboring deserts. Annie Alexander, who founded the Museum of Vertebrate Zoology at the University of California at Berkeley, trapped mammals in the Tularosa in 1931 with Louise Kellogg, her life partner and fellow biologist. Many others followed suit, making the Tularosa one of the most intensively studied areas in the American Southwest. The findings from their surveys were at first difficult to interpret. There were legions of animal species in the valley, but only some of them showed an obvious color correspondence to either black lava or white gypsum sand. Of the eleven rodent species that were trapped from the white dunes, only four were lighter colored than their relatives in the surrounding desert. Although the dune-living pocket mice and woodrats were pale in comparison to their relatives outside, other dune species, such as kangaroo rats, showed no such distinction. Similarly, of the seven rodents abundant on the black lava, only three were darker than their nearby relatives. The same was evident for reptiles: Of the seven genera of snakes and lizards on lava outcrops in the Tularosa and northern Mexico, just four were relatively dark.

The accepted explanation had to do with mobility. A large part of the coloration differences among these species was apparently a reflection of differences in their degree of isolation. Some species move around a lot, but others tend to stay put, spending most of their lives in a specific, uniformly colored habitat and rarely venturing into areas with contrasting backgrounds. Most of the isolated species in the surveys showed a clear resemblance to the color of the background, but none of the broadly distributed ones did.[9]

Isolation is a powerful evolutionary phenomenon. In the Tularosa, the more isolated desert species, those limited to relatively small and specific habitat areas, mate only with other individuals in that same area. Like people in a remote village, these animals exchange genes within a small population, many of whom share the same genetic traits. Over time, these isolated populations begin to share characteristics that are different from populations in other areas. Discovery of the black and white forms of animals in the Tularosa provided a clue as to why some animals match their color environments and others do not, and the key was in what happens to isolated populations.

8. Flowing Genes

WHAT DOES THE COLOR of lizards and pocket mice have to do with dodos, giant flightless moas, and pygmy mammoths? Each is the result of evolutionary processes that occur when populations of animals are isolated. Some of the best places to study the effects of isolation are archipelagos, chains of islands such as Indonesia or the Galápagos Islands, which inspired Alfred Russel Wallace and Charles Darwin to independently formulate their theories of evolution by natural selection. Wallace spent eight formative years in the Malay Archipelago, which includes modern-day Indonesia, the Philippines, Singapore, Brunei, and eastern Malaysia. Darwin was directly inspired by his brief contact with the Galápagos: When he returned from his five-year voyage as a naturalist on the HMS *Beagle*, he sorted out the specimens he had collected there and was astonished to discover that each of the islands was effectively its own little world. "I have not as yet noticed by far the most remarkable feature in the natural history of this archipelago," Darwin wrote, "it is, that the different islands to a considerable extent are inhabited by a different set of beings" (Darwin 1845, 393–94).[1]

Islands have long been considered cradles of evolution, but they don't always have to be surrounded by water. The evolutionary consequences of island isolation occur regardless of whether an animal population is in the middle of the ocean or on a rocky outcrop in the desert. Ernst Mayr, a curator from the American Museum of Natural History, spent much of his career studying birds from the archipelagos of the South Pacific. Across the islands, there were many different types of monarch flycatcher, with some distinctive forms on islands separated by less than ten

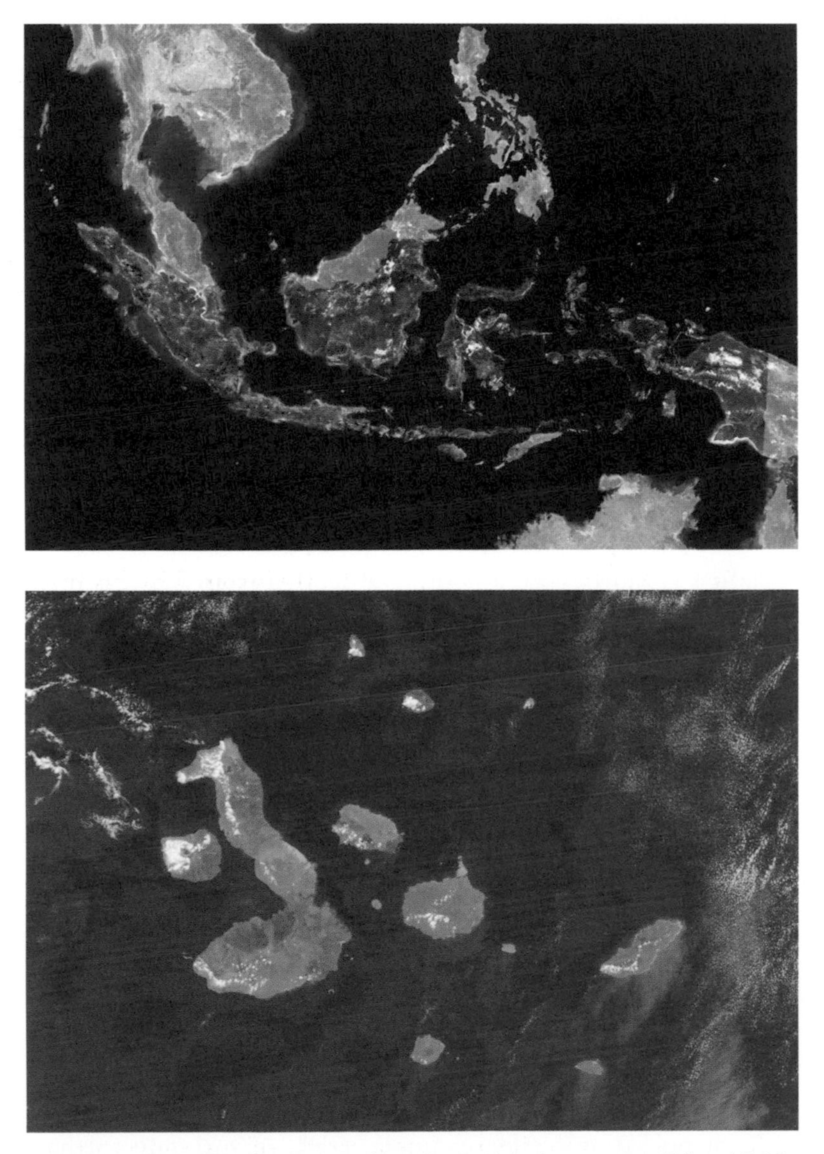

Archipelagos are chains of islands often created from the same volcanic source. **ABOVE:** Wallace visited the Malay Archipelago, shown in a satellite photograph. Located in Southeastern Asia, it contains over twenty thousand islands. Photograph by Planet Observer (Getty Images). **BELOW:** Darwin visited the Galápagos Islands off the coast of Ecuador, consisting of thirteen main volcanic islands and numerous smaller ones. NASA GSFC image courtesy Jacques Descloitres, LANCE/EOSDIS MODIS Rapid Response Team.

kilometers of ocean. The conspicuous geographic variation in the form and color of these birds convinced him that he was seeing on these islands the primary engine of the evolution of biological diversity.

Islands were central to Mayr's understanding of evolution, and the key feature was isolation. He began his argument by noting that variability is pervasive in the biological world. Careful measurement, even within a single species, invariably reveals that each individual is unique, carrying its own distinctive combination of genetic traits. Mayr went on to show that features from one population of a species are also consistently different from those of more distant populations, and the greater the separation between populations, the greater the divergence. This pattern of variation is shown even among mainland species, but it is most conspicuous on island chains.

Mayr noted that it is not necessary for a piece of land to be surrounded by water to function as an island. The San Francisco Peaks, where Merriam collected his rodents, could be considered habitat islands because their high-altitude forests of pines and firs are isolated from other mountain areas by wide expanses of lowland desert, where few forest animals can survive. For some species, deserts provide a mosaic of contrasting environments at an even smaller scale. Dune fields or rock outcrops that are separated from similar regions by even a few kilometers of hard, crusted clay or gravel are also effectively habitat islands, at least for animals that strongly favor rocky cliffs or loose sand to make their living.

Rock pocket mice, for example, stick very closely to their jumbled piles, choosing to burrow only under large boulders, and fringe-toed lizards are seldom seen farther than fifty meters from a patch of wind-driven sand dunes. These animals can be every bit as isolated from other members of their species as the butterflies, finches, and tortoises that Wallace and Darwin studied on their island chains. In Mayr's view, the most rapid evolutionary innovation takes place in such fragmented habitats, among small, isolated populations. Evolution does occur in large, fully interbreeding populations, but it happens more slowly and incrementally. Unusual animals such as dodos, moas, and pygmy mammoths evolve on islands as a consequence of their separation from mainland relatives.[2]

Mayr's explanation began with the genes, the basic currency of inheritance. Genes are the DNA tokens that are exchanged and recombined during reproduction and that code for each individual's constellation of traits. The ultimate source of all genetic variation is mutation, the occasional, spontaneous alterations in the DNA code that can result in different forms of the same gene. Evolution, at its most basic level, is the result of changes in gene frequency, modifications in the relative abundance of the various forms of a gene. The driving force in evolution is differential reproduction—the sorting out of which individuals survive long enough to breed, who breeds with whom, and which individuals actually succeed in raising young. In Mayr's view, three population-level processes account for most evolutionary change. The first is genetic drift, a fluctuation in gene frequencies due to chance. This occurs, for example, when a development project separates two populations of snakes or a river changes its course and divides two populations of mice. Just by chance, the separated populations can start off with different gene frequencies and evolve in different directions.

The second process is gene flow, a change in gene frequencies in a population due to animals migrating in and breeding with the locals. Think of it as a stream of genes trickling in from outside and blending with the ingredients already present in the population. Gene flow can introduce new genetic variability into a population and provide a powerful counterbalance to inbreeding. The third process is natural selection, in which factors in the environment—such as the presence of predators or the availability of food and water—change the odds of survival and reproduction for individuals with particular sets of traits. Traits that promote success in particular environments become more common in the population over time. In natural selection, Darwin noted, "[E]ach new variety, and ultimately each new species, is produced and maintained by having some advantage over those with which it comes into competition; and the consequent extinction of the less-favoured forms almost inevitably follows" (Darwin 1859, 320).

The processes of genetic drift, gene flow, and natural selection are often studied in small populations, where their independent effects are easier to see, but every natural population is influenced by all three. Even continental species are now seen as aggregates of small populations, separated to at least some degree by distance and by evolved preferences for their local habitats. Against the background of genetic drift, the evolution of particular traits reflects both natural selection within local populations and gene flow among them. When gene flow is restricted, small geographical differences in the direction and intensity of natural selection can cause local populations to diverge from one another. The association between the degree of genetic isolation and the occurrence of local forms that closely match the hue of their backgrounds is one of the strongest indications of the role of natural selection in the evolution of concealing coloration.[3]

What if you could study exactly how genes flowed among populations of differently colored animals? Scientists test these ideas by quantifying the resemblance between animals and their local background, determining which genes are responsible for the concealing coloration and then measuring the overall degree of relatedness to determine the gene flow between populations. It is not a simple undertaking, but innovative researchers have employed the tools of molecular biology to examine selection and gene flow in two groups of desert animals. In the process, they have begun to shape a modern understanding of the mechanism by which matching coloration evolves.

Rock Pocket Mice

With their round faces, small ears, and long furry tails, pocket mice look a bit like gerbils. They forage for seeds, gathering them up and carrying them back to granaries in their burrow systems. Pocket mice have external, fur-lined pouches on their cheeks, which they stuff full of seeds, holding them in by a ring of muscles. When the mice return home, they turn their pockets inside out and brush out the seeds, the way you would clean dryer lint out of your pants. Pocket mice forage nearly every night when seeds are abundant, leaving networks of trails in all directions. When they are out in the open, pocket mice are attacked by many nocturnal predators, particularly owls, but also western rattlesnakes, coyotes, badgers, and ring-tailed cats. Only about one pocket mouse in six survives for even a single year.[4]

One pocket mouse species, the rock pocket mouse, has an additional crucial aspect to its biology: They have an almost obsessive attachment to rock outcrops. These little rodents cling to their craggy sanctuaries like barnacles on a rocky coastline. Rock outcrops are sparsely distributed across the desert Southwest, and they occur as a clump of granite boulders at one location and a cinder cone or small lava field at another, more distant point. The space between them is usually sand or desert hardpan, wholly unsuitable for picky pocket mice. Because of their strong preference for rocks, these mice are effectively isolated on Mayr's habitat islands, where they experience reduced gene flow from neighboring populations. They differ strikingly in coloration from one outcrop to another, their coloring closely corresponding to the local geology. Over most of their range, they live on creamy to reddish-orange boulder piles, and they have a light, sandy-colored pelt. But on lava outcrops like the Carrizozo Malpais, their fur is dark brown to coal black.[5]

Lee Dice, from the University of Michigan, and his colleague, Philip Blossom, conducted the first quantitative study of pocket-mouse coloration—one of the first such studies of any species—as part of their survey of the mammals of the desert Southwest. Beginning in 1930, they systematically observed and collected all sorts of mammals, from bats to skunks to mule deer, across a wide swath of southern Arizona and New Mexico, but the centerpiece of their fieldwork was the rock pocket mouse. They trapped 335 pocket mice at seventeen locations, covering the full range of rock coloration, and then analyzed the reflectance of the mouse skins and of the rocks near where each mouse was captured. Dice also collected experimental evidence indicating that mouse coloration in general makes a difference, demonstrating that owls had a more difficult time detecting and capturing deer mice that match the coloration of their backgrounds.

Hopi Hoekstra, an evolutionary biologist from Harvard University, and her colleague, Michael Nachman from the University of Arizona, realized that the rock pocket mouse offered an extraordinary opportunity to investigate the effects of isolation on animal coloration using modern molecular genetics. One

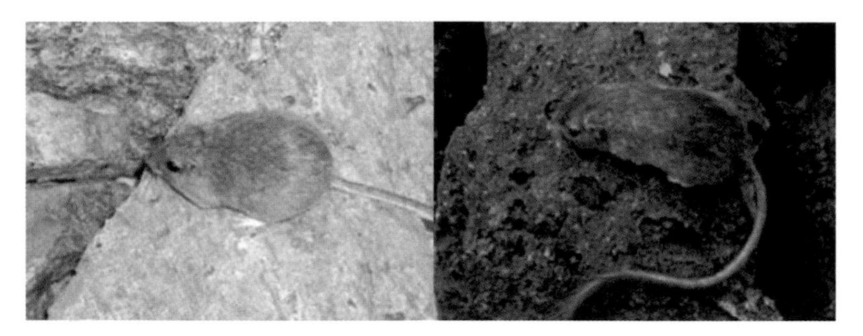

Isolated populations of rock pocket mice (*Chaetodipus intermedius*) display dramatic coloration variations. **LEFT:** The light form of the species. **RIGHT:** The dark form from the Carrizozo Malpais. Photographs by Michael Nachman.

of their studies was conducted among rocky outcrops in southern Arizona. Their sites included the black Pinacate lava beds and several ridges of lighter colored granite, all separated from one another by kilometers of pale yellow sand. Hoekstra and Nachman collected several hundred rock pocket mice from these areas and painstakingly measured how closely the color of the mice matched the color of their substrate. The researchers then sequenced the gene that determined differences in mouse coloration. Hair color in mammals is entirely the result of two kinds of melanin pigments: eumelanin, which produces dark brown or black hues, and phaeomelanin, which generates blond to auburn. Coat color differences in many mammals result from mutations in a gene called *Mc1r,* which regulates the synthesis of the dark eumelanin. Some mutations in *Mc1r* can allow eumelanin synthesis to run wild, producing a much darker coat color. Hoekstra and Nachman found that the dark mice from the Pinacate lava beds had a mutant form of *Mc1r,* which caused a greater production of eumelanin.[6]

They then set about the task of analyzing the population genetics of their mice. The most reliable way to measure gene flow is to find out how the individuals in different populations are related to one another. Hoekstra and Nachman used DNA sequencing to calculate the genealogies of 175 adult mice. From this, they estimated the number of breeding females in each population and the proportion of those females that had migrated in from elsewhere. The lava bed and one of the granite sites were separated by ten kilometers of sand. The breeding populations were about 10,000 females apiece, but in spite of their proximity, they exchanged no more than about 80 individuals per generation. So the first conclusion was that the two outcrops were definitely desert islands, with clear indications of restricted gene flow. The researchers also found that the gene frequencies went along with the backgrounds: The dark *Mc1r* gene was found in 86 percent of the mice on the dark lava, while 66 percent of the mice on light granite had the light form of the gene. Hoekstra and Nachman concluded that natural selection was acting on the color of the mouse populations.

The balance of gene flow and selection did not work the same way on the two colors of mice. More light mice survived as they made the dangerous journey across the open space between outcrops. Dark mice crossing the pale yellow sand between the outcrops stood out from the background and were more likely to be picked off by predators. The sand was less of a barrier for light mice, and as a result, the light mice had a higher rate of gene flow.

Even after the mice reached their rocky outcrops, predators didn't act equally on the light and dark individuals. Predators found light mice on the dark lava more easily than dark mice on the light granite rocks. Natural selection against conspicuous light mice on dark lava is fierce: Their survival rate is about a third less than dark mice on the lava beds. Thus the light mice have higher rates of gene flow but experience stronger natural selection. For rock pocket mice, natural selection and gene flow are powerful forces that drive the evolution of background matching, but there is a tilt to the system, with light and dark mice evolving at different rates.[7]

The story of how isolation affects the evolution of coloration doesn't end with pocket mice. Another piece of evidence comes from a set of modern studies conducted on the lizards of the white gypsum sands of the Tularosa Basin.

Lizards of the Tularosa

Lizards derive their body heat from the external environment, unlike birds and mammals, which run their own metabolic furnaces. In cold regions, this is a disadvantage, since cool temperatures limit the time that reptiles can actively move around. Hot deserts are a paradise for lizards, and most times of the day are perfect for foraging and mating. The same deserts that are home to seed-eating rodents during the cooler night hours are overrun with lizards during the day. Desert insects feed on leaf litter, seeds, and other detritus, and the lizards feed on the insects, playing a pivotal role in the ecology of the community.

The pale gypsum dunes of White Sands National Monument are geologically recent. They reached their current form less than five thousand years ago. Before the dunes were deposited, the Tularosa Basin was typical Chihuahuan Desert—light reddish-brown sand and gravel with scattered clumps of mesquite, agave, and bunch grasses. This is how the central part of the basin still looks today. Bree Rosenblum, a biologist from the University of California at Berkeley, surveyed the central basin and found that it hosts some eleven different lizard species. When she surveyed the white sands, she found only three: eastern fence lizards, lesser earless lizards, and little striped whiptails.

Rosenblum and her colleagues reasoned that the arrival of the dunes must have provided an ecological opportunity for the resident lizard fauna. The white dunes were an unexploited habitat, offering an abundance of insects and few competitors. The only problem was that lizards throughout the Tularosa are a major source of food for predatory birds, as well as a range of snakes, other lizards, and the occasional coyote. So far, only three of the various species of lizards that live in the Tularosa Basin have evolved pale color forms, allowing them to make a living in the white sands without being conspicuous to predators.[8]

All three lizard species occur in a pale form on the white gypsum dunes and a darker, more typical form on the adjacent red-brown sand. To understand the balance between natural selection and gene flow, Rosenblum first quantified how well her lizards matched the colors of their local soils. She then determined the genetic relationships among individuals in each environment, which allowed her to calculate gene flow. Finally, she estimated the relative impact of natural selection and gene flow. This was a similar research approach to the one adopted by Hoekstra and Nachman, but Rosenblum's lizard study had two novel features. First, the Tularosa dunes had three lizard species responding to the same environmental challenge, so Rosenblum could compare them to see what determined the differences in their color evolution. The second feature was that these animals were not physically isolated, since the differently colored sands were right next to each other and the lizards could easily shift back and forth between them. If gene flow was limited between these lizard populations, it was not because of physical isolation.

Rosenblum found large differences among the three lizard species in how well their coloration matched the substrate.

ABOVE: The lesser earless lizard (*Holbrookia maculata*) shows the most extreme variation in coloration of any lizard species in the Tularosa. Photograph by Simone Des Roches. **BELOW:** Coloration differences among populations of little striped whiptails (*Aspidoscelis inornata*) are also striking. Photograph by Erica Bree Rosenblum.

Although all three species were lighter in the dunes than on the dark soil nearby, the lesser earless lizards on the dunes offered a better match to their background than either of the other two species. They also showed the greatest difference in color between dark soil and white sand forms. In all three lizard species, there were mutations of the *Mc1r* gene that were associated with the loss of coloration, though they were all different mutations that affected melanin regulation in strikingly different ways. Rosenblum analyzed the color differences across the gradient from white to dark and related them to her measure of gene flow. To determine levels of isolation between white sand and dark soil populations of the lizards, she sequenced the mitochondrial *ND4* gene and associated tRNAs. Gene flow for the three different lizard species occurred at different rates. Earless lizards were more sedentary, showing the lowest rate of gene flow, fence lizards showed an intermediate rate, and whiptail genes traveled freely across the boundary in both directions. So the species with the greatest color resemblance was the one that showed the strongest evidence of genetic isolation.

There is one more twist to the story. In the case of the rock pocket mice, physical barriers separated the populations, but the lizards experienced a different kind of barrier. They were not prevented from moving between dark- and light-colored habitats, but their gene flow was restricted by their breeding behavior. Mice have no problem breeding with partners with different hair colors. Like most rodents, they recognize an appropriate mate mainly by smell. But lizard breeding is all about color and movement. Male lizards have bright coloration on their heads or sides that is very conspicuous when they perform their courtship displays. Female lizards also employ color cues to indicate their sexual receptivity. Rosenblum and her students showed that the same genetic changes that make the dune lizard populations less vulnerable to predation also modified the colors of their mating signals. The result is that the whitened animals are less able to breed with darker individuals. The process has gone furthest with lesser earless lizards, which now show distinctive preferences for mating with individuals of their own color, suggesting that the light and dark forms are on the way to becoming distinct species.[9]

Dodos, moas, and pygmy mammoths all evolved on islands that were physically isolated from their mainland populations. Over time, the combined effects of genetic drift reduced gene flow from mainland populations, and natural selection shaped the evolution of these unusual species. Isolation has proved to be a key component in the evolution of concealing coloration, as well, even in the absence of water barriers. It is the fulcrum in an evolutionary seesaw that balances gene flow and natural selection. But isolation can take many forms. A stretch of sand can isolate rock pocket mice that cling to rocky outcrops, and when natural selection results in a trait like pale coloration in lizards, it can be accompanied by mating preferences that, in turn, serve as further isolating mechanisms.

Oceanic islands informed the evolutionary theories of Wallace, Darwin, and, eventually, Mayr. But for understanding the evolution of concealing coloration, deserts provided the clearest data. Deserts reveal the closest relationships between the colors of animals and the colors of their environment. Nearly everywhere else, with the exception of polar regions and the deep sea, plants provide the dominant background against which animals are exposed to visual predators. But in deserts, plants are thinly spread, and the underlying geology and the geochemical processes that tint desert landscapes are relatively unobscured. Animals play out their evolutionary dynamics against a geological background of rocks, sand, and soil.

The seamless match between a desert animal's coloration and its background was noticed by Napoleon's young naturalists. It fascinated Tristram in his wanderings, and it has aroused the curiosity of scientists ever since. Their observations are now understood in light of what we know about the combined influence of environment, gene flow, and natural selection. Against the naked environment where desert animals spend much of their lives, small differences have large impacts. Where a desert stretches as a uniform blanket of sand, unrelated animals have a similar bland color. But deserts can also be places of alternating patches of contrastingly colored habitat with abrupt changes from one to the next. These patches set the stage for the evolution of distinctive color patterns.

The degree to which animals confine themselves to a particular habitat affects gene flow among their populations. The degree of exposure to predators and competitors in its habitat and the qualities of the predators' visual abilities determine the strength of natural selection. Where natural selection is strong and gene flow is limited, animals can evolve color patterns that match their local habitat patches. These effects can be so powerful that individual populations of the same species can evolve into the black and white, the yin and yang of contrasting coloration.

Part Four

DETECTION

[T]he Giraffe grew blotchy, and the Zebra grew stripy, and the Eland and the Koodoo grew darker, with little wavy grey lines on their backs like bark on a tree trunk; and so, though you could hear them and smell them, you could very seldom see them, and then only when you knew precisely where to look.

RUDYARD KIPLING, "How the Leopard Got Its Spots," *Just So Stories*

Blue jays (*Cyanocitta cristata*) learn to distinguish minute characteristics of their prey. This jay has been trained to find images of moths on a computer screen, helping scientists at the University of Nebraska understand the process of searching.

9. Telltale Signs

ALONG LA RAMBLA, the famous boulevard in Barcelona, painted actors pretend to be statues. They are quite believable, until the statue takes a drink or asks for a donation. If we can be fooled, how do other animals tell the difference between real and pretend?

Visual predators live in a world of deceit. Dedicated imitators, with their lives at stake, do practically anything to avoid detection. The elements of animal deceit include nearly every aspect of identity: how an animal looks, smells, sounds, and behaves. Does it look like a shrimp or a frond of seaweed? Is it a caterpillar or a branch? A mantid or a stone? The cues that distinguish real prey are often exceedingly subtle, but with sufficient experience, predators can learn them. Learning is, in fact, the primary advantage that visual predators possess in their endless struggle to locate camouflaged prey. Some of the most impressive evidence of predatory expertise comes from studies in tropical rainforests, where concealing coloration is a nearly universal strategy.

Tamarins are the smallest monkeys in the New World, roughly the size of a young squirrel. They live among the tangled vines and branches of the rainforest understory, where they hunt for many kinds of food, particularly grasshoppers and katydids, but also beetles, spiders, caterpillars, snails, small lizards, nestling birds, and sometimes fruit and tree sap. Tamarins are dedicated visual predators: More than a third of their daily activity is spent in carefully scanning branches and leaves for any signs of prey. When they spot a possible target, they stare at it fixedly, apparently deciding whether it is edible, and when they have made their decision, they pounce, often from more than a meter away. The prey may occasionally escape, but experienced tamarins rarely make mistakes in recognizing an appropriate target. The cues used to discriminate valuable prey from useless foliage are apparently sufficiently reliable that tamarins do not pounce on sticks or leaves or empty branches.[1]

The Smithsonian Institution operates a research station on an island in Panama dedicated to the investigation of tropical biology. Michael Robinson spent two decades there researching concealing coloration in rainforest insects. At Oxford University, he had been a student of Niko Tinbergen, one of the founders of the experimental study of animal behavior, and Robinson came to share his advisor's fascination with camouflage and visual search. He was particularly interested in how tamarins are able to recognize camouflaged prey. Robinson noticed that the monkeys invariably bit insects in the head before consuming them, and he wondered whether they might be especially attuned to head structures. To investigate the cues that they were using, Robinson brought wild-caught red-crested tamarins into the laboratory and tested them in large aviaries.[2]

In his first experiment, Robinson scattered sticks on the floor of the aviary and noted that the monkeys showed no interest in them. He then glued the head of a cricket on one of the sticks, placed it among the others, and reintroduced each tamarin for a period of foraging. Of the thirteen monkeys he tested, nine discovered the modified stick, seized it, and bit at the cricket head. After eating the head, the monkeys then bit at the stick, but eventually dropped it when it proved not to be edible. The tamarins appeared to recognize insect heads as a sign of the presence of a prey item.

Cricket heads mounted on sticks are not particularly natural stimuli, however. Robinson's later experiments used phasmids,

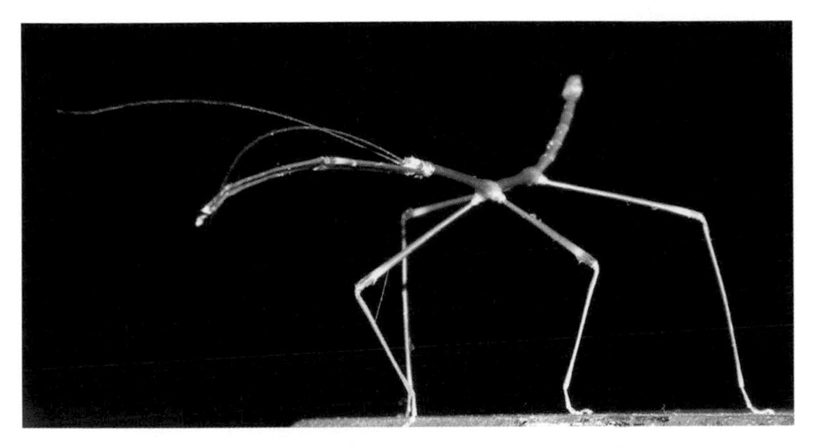

This phasmid from the Tabin Wildlife Reserve in Borneo stands with legs extended. The two forelegs are held together near the antennae, hiding the head. Photograph by Judy Diamond.

or stick insects, which have strikingly elongated bodies and thin, branchlike legs. Most phasmids stand at rest with their middle and hind legs extended out to the sides and their forelegs brought forward, concealing the head. Robinson felt that these insects would be more of a challenge for head-hunting tamarins, though their extended legs might still provide a cue to their presence. He tested this idea with two local Panamanian stick insects that drop from their resting places when disturbed and assume what Robinson called a "cataleptic stick attitude." Basically, the insects bury their heads between their forelegs and fold down the middle and hind legs along the sides of the body, greatly enhancing their stick resemblance.

Robinson prepared dead phasmids in various leg postures, either with the legs concealed next to the body or with one pair extended in a typical walking pose. He then gave tamarins an opportunity to search for these modified insects under four conditions: legs concealed on the floor among other sticks, legs concealed on the floor without sticks, legs concealed on a branch

The red-crested tamarin (*Saguinus geoffroyi*) searches for phasmids among sticks in Robinson's experiment. There are two phasmids in the picture, one with legs extended and one with legs concealed. Illustration by Angie Fox.

in the aviary, or legs extended either on the floor or on a branch. The results were unequivocal. Whether there were additional sticks on the floor or not, the monkeys did not discover phasmids with concealed legs, even when they were given half an hour to search. But when the phasmid legs were extended, tamarins found and consumed them within two to three minutes of active searching. The outcome suggests that even very experienced and perceptive predators can be fooled by a sufficient resemblance to an inedible object, but that any small, consistently insectile cue is likely to destroy the illusion and leave the phasmid open to attack.[3]

How do animals acquire this perceptual expertise? To begin with, tamarins and their relatives, the marmosets, are masters of discrimination. A common psychological learning task involves giving subjects pairs of distinctive objects with a reward hidden inside one of them. Subjects may be given hundreds of these matched pairs in order to measure their ability to remember and discriminate. In the laboratory, common marmosets quickly learn up to a thousand discriminations over a period of days and recall them later with perfect fidelity. In the wild, marmosets and tamarins learn to differentiate the valuable from the useless, and they are particularly good at differentiating items with multiple dimensions and complex shapes.

Many visual predators are opportunistic about what constitutes prey. On their own, young animals readily approach novel items, and they often attempt to eat whatever they come across, including things that are clearly inedible or even dangerous. Keas, which are omnivorous New Zealand parrots, are particularly open-minded when they are newly out of the nest. On a backcountry rubbish dump, we observed fledgling keas make such unfortunate dietary choices as string, masking tape, flashlight batteries, and foam rubber, and when the dump was burning, they even tried to bite the flames. But the foraging behavior of parents or other related adults can help to guide and focus juvenile explorations. Young keas are drawn to the food choices of older birds, and over time, they come to discriminate acceptable food from inedible objects.

As in keas, the acquisition of foraging in tamarins and marmosets is influenced by the behavior of older individuals. Field studies of common marmosets found that infants spend half their foraging time riveted on the behavior of the adults. Older juveniles do not watch adults as obsessively, but most of their foraging behaviors are direct copies of those of their parents. In open aviaries, family groups of marmosets are highly attuned to small changes in their environment, and they immediately approach and examine anything that is even remotely novel. In many cases, the new item is merely sniffed or touched by one member of a family. If it is not worthy of further interest, it is not approached again by other members of the group. If the object provides food, however, the finder gives a food call, alerting its relatives. Once an object has been linked to food, it is persistently investigated by group members on a regular basis. This combination of compulsive exploration, rapid foraging decisions, and a long memory for rewards has been termed the "win-stay, lose-shift" strategy. It is an optimal approach for an active, generalist predator, as it maximizes the chances of finding new food sources while minimizing the amount of time spent fiddling around with rocks, sticks, and leaves.[4]

The studies on tamarins and marmosets provide insights, but they don't tell the whole story of how animals learn to find camouflaged insects. Robinson's monkeys were captured in the field, so they were already experts at finding rainforest insects, and psychological studies on lab-reared marmosets never asked them to discriminate anything as difficult as twigs and stick insects. Marmosets are equipped with remarkable cognitive tools, but we don't know how much experience is really required to become a professional insect hunter. One experiment did address this issue, but instead of primates, the researchers examined how Eurasian jays find camouflaged caterpillars.

Eurasian jays are forest birds, inhabitants of a great swath of deciduous woodlands from Western Europe and the Middle East across southern Russia to China and Japan. Their primary staple is acorns, which they gather up in quantity in the fall and cache throughout their territories. Eurasian jays are omnivores that feed on everything from fruit and grain to mice and lizards. They are notorious nest robbers, consuming both eggs and young birds, and expert predators, hunting both in trees and along the ground for caterpillars, beetle grubs, and other invertebrates. In the 1950s, the same Dutch scientist who studied countershading in caterpillars, Leendert de Ruiter, and his advisor, Niko Tinbergen, examined the foraging behavior of Eurasian jays. De Ruiter tested hand-raised jays that were learning to search for stick caterpillars, and he used caterpillars of geometrid moths, which bear a marked resemblance to twigs of their usual food plants, imitating both the color and the texture of the bark.

The camouflage of stick caterpillars goes well beyond coloration: They bear lumps on their body surface that resemble buds or leaf scars, and they adopt a rigid posture that enhances their twig resemblance, gripping a branch with the clasping legs on the end of their abdomen and holding the body out at an angle. The caterpillars remain motionless in this pose throughout the day; they only relax and begin feeding at night.

De Ruiter obtained caterpillars from the wild and harvested sticks of comparable size from the trees where they lived. For each experiment, he scattered several recently killed caterpillars among similar sticks on the floor of an aviary. He then introduced eight hand-reared jays, one at a time, and recorded their foraging behavior. The jays had been accustomed to being fed in the aviary and were used to sticks on the floor, but they had never previously encountered stick caterpillars. They were hungry, so they scoured the area for possible prey items. Even so, it took them an average of nearly half an hour to find their first caterpillar. A common initial contact was stepping on a twig and finding it squishy. Immediately after their original discovery, however, half the birds began frantically pecking sticks all over the aviary. All it took was one caterpillar encounter to get them excited about chewing on

De Ruiter studied how stick caterpillars, such as this geometrid moth, are recognized by bird predators.
Photograph by Ingo Arndt/Foto Natura (Getty Images).

Eurasian jays (*Garrulus glandarius*) prey on stick caterpillars. Photograph by Peter Lilja (Getty Images).

sticks. Because there were other caterpillars in the aviary, these intensified stick attacks were occasionally rewarded with additional successes, helping to sustain the behavior. Their response to the first discovery indicates that the jays had overlooked the caterpillars at the beginning, not just because the prey were unfamiliar but because the birds were actually unable to distinguish them from sticks.

The remaining jays were more cautious. Although they also found their first caterpillar through an accidental encounter, they subsequently pecked very few sticks. Instead, they meticulously examined each object on the aviary floor and slowly, systematically picked up one caterpillar after another. The implication is that they had learned something in their initial discovery, some telltale sign of caterpillar identity, and they spent the rest of their experimental session focusing exclusively on that feature.

Other species of jays have been shown to have similar learning abilities. Blue jays are members of the North American branch of the family. Like Eurasian jays, blue jays are reliant on acorns, but are otherwise omnivorous. They are similarly expert predators, consuming a wide range of invertebrates, as well as small birds and mice. Blue jays learn new visual discriminations rapidly and retain them for long periods, following a "win-stay, lose-shift" strategy. They are phenomenally sensitive to small, consistent visual cues. Blue jays are one of the few North American birds that can find camouflaged moths on tree trunks, and they readily learn to detect them in photographic slides of natural bark backgrounds. In discrimination experiments using digitized moth images on complex, granular backgrounds, blue jays are consistently able to locate cryptically colored moths that are nearly invisible to humans.[5]

That jays and tamarins can learn to find stick mimics is evidence that no coloration strategy, not even elaborate forms of object resemblance, is proof against the efforts of specialized visual predators. But the testing conditions in these studies were somewhat biased against the insects. In both Robinson's and de Ruiter's experiments, the predators were confined to an aviary in which they had been accustomed to finding food, which ensured that they would invest a lot of effort in searching the area. Plus, the insects were not displayed in their typical locations and postures, which may have cut into the effectiveness of their disguise. Probably the most important difference between these experiments and the natural environment, however, is that no other food items were offered. Concealing coloration works, not by providing a perfect simulation of the surroundings but by making the prey species just enough harder to find that predators become discouraged and search for something else. If the tamarins or the jays had other, more obvious food available, it is unlikely that they would have discovered these highly camouflaged organisms.

Niko Tinbergen contended that there are four adaptive behavior strategies that prey use to deceive predators. The first is immobility. Motion is highly salient to visual predators and draws their immediate attention, so many camouflaged insects move around only at night. The second is background selection. Stick mimics are sensitive to discrepancies in the appearance of the surrounding shrubbery and will move away from nonmatching substrates. The third behavior is the adoption of special postures that maximize their mimetic resemblance: Panamanian phasmids pull their legs in close to the body, and European caterpillars hold themselves rigid at an angle to an underlying branch.

It is Tinbergen's fourth criterion that is somewhat surprising. Highly cryptic animals, he contended, should be locally uncommon, and the animals that use this strategy should tend to be widely scattered across the habitat. This prediction appears to be true, at least in Central America. Robinson noted that in any particular habitat, there always seem to be fewer stick and leaf mimics than there are insects with simpler, more generalized types of camouflage. Basically, object resemblance can't be maintained when the prey are common. When a stick or leaf mimic increases in numbers, visual predators begin to acquire more experience at recognizing it, and eventually this will limit the size of the local population of the mimic. Prey that are so closely spaced that their presence teaches predators the appropriate discriminative cues will be locally exterminated, leaving only a few individual prey distributed through the habitat. By Tinbergen's criteria, camouflaged prey should be motionless, discriminating in their habitat choice, deceitful in their behavior, and rare.[6]

Natural objects tend to be variable in appearance. Think of how many different shapes and colors of leaves there are in forest litter or how varied the twigs are on a single tree. This abundance of model forms works to the benefit of camouflaged prey. Even a partial resemblance to natural objects can have significant evolutionary benefits. For every inventive and deceptive strategy used by their prey, visual predators eventually learn the small consistent discrepancies that indicate edibility. Predators experienced with both real sticks and stick mimics learn the telltale cues that distinguish the edible from the useless, the real from the make-believe.

10. Psychology of Search

IN IAN FLEMING'S short story "The Hildebrand Rarity," James Bond is vacationing in the Seychelles Islands when he is given the task of collecting one of the world's rarest fish. It has only been seen once before, so all Bond has to go on is a verbal description of a pink- and black-striped fish with sharp, spiny fins. As he swims along the reef, Agent 007 muses to himself, "When you are looking for one particular species under water,... you have to keep your brain and your eyes focused for that one individual pattern. The riot of colour and movement and the endless variety of light and shadows fight your concentration all the time" (Fleming 1960, 312). Fleming was an experienced birdwatcher, so he was familiar with the difficulties of visual search in a natural environment. But the task Bond faced is in some ways familiar to everyone. Think of having to find the car keys you misplaced somewhere in the house or the book you need to return to the library or looking for a particular piece in a jigsaw puzzle. In each case, you focus on specific features of the sought-for object. Perhaps the key ring is decorated with a bright orange talisman, or maybe you remember that the library book is blue with gold lettering, or perhaps the puzzle piece has a lot of green foliage.

What we see through our eyes appears to be an accurate and unbiased image of our surroundings, but that is an illusion. Our view of the world is an elaborate cognitive reconstruction that is continuously modified to suit our changing needs. By focusing your visual search, you actually change the appearance of your environment to emphasize the things you are searching for. The bright orange key ring may appear to pop out and hover over the surrounding clutter, while anything that is not orange fades into a blur of insignificance. Or your eyes will feel drawn to each of the green puzzle pieces, ignoring all the others.

If you are looking for something specific, a focused search is vastly more efficient than an undirected scan of the entire environment. That is especially true if the environment, like Bond's coral reef, is a confusing pastiche of contrasting colors and textures. The focusing process is called visual attention, and it is one of our most powerful cognitive abilities. It converts the visual system from a general processor of all incoming information to a dedicated search engine that is concentrated solely on the most relevant parts of the visual field. Attention makes it much more likely that you will find what you are looking for, but it also has a down side: As a rule, you will fail to find what you are *not* looking for. If your recollection of your library book is incorrect—if, for example, it is red, instead of blue, with black lettering, instead of gold—your focused search is doomed, and you would have been better off systematically examining each book in the house, one by one, without preconceptions.[1]

The concept of a searching image to explain the role of attention in visual search was the invention of Jakob von Uexküll, a German biologist and professor at the University of Hamburg. In his best-known story from 1934, von Uexküll wrote that he was used to having water served at meals in an earthenware pitcher. One day, the pitcher broke, so his friend replaced it with a glass carafe. When von Uexküll came to the table, he looked in vain for the pitcher. He asked for water and was told that it was in its usual place. Only then, he claimed, did the scattered reflections from the glass carafe "fly together" and form themselves into a coherent image. Von Uexküll asserted that the expected visual features of the pitcher overrode the perceptual input and prevented him from recognizing the carafe.

Visual attention is a very flexible capability. You can focus on particular colors or patterns, on individual spatial locations, or on a combination of features, such as a puzzle piece with a specified color and shape. Or you can conduct a categorical search, looking for, say, all the vehicles in a photograph or all the nouns on a page of text. These are examples of a guided search that is directed from the top down by processes at higher levels in the nervous system. Top-down guidance may be the most common way to think about visual attention, but it is not the only way that it works. The focus can also come from the bottom up, from aspects of the sought-for object and the external environment. In naturally occurring bottom-up searches, an image from the immediate past is the most influential guide. Predators remember the most recent prey item they discovered, retaining a residual impression in their short-term memory. When they encounter another prey that matches this searching image, it is more readily recognized. Where the visual input does not match the searching image, prey are often overlooked.

People make ample use of searching images. One unexpected context is sorting. Suppose you have a bag of small hardware—screws, nails, nuts, bolts, and so on—and you decide to organize them into little jars. You dump the stuff out on a table and begin separating the items into coherent groups. It is possible to do this by randomly picking up individual objects, one by one, identifying each one, and then moving it to the appropriate jar. But what most people do is very different. They quickly pick out a whole series of items of the same type, making a handful of, say, small screws. They put them in the jar and then go back and do the same for a different kind of item. So the sorting sequence is nonrandom, producing runs of items of a single type. It is a faster, more efficient technique, and much of the increased efficiency is due to the use of searching images.[2]

The connection between searching images and visual predation was first proposed in 1960 by the Dutch ecologist Luuk Tinbergen. While studying the prey items that insect-eating birds feed to their young, Tinbergen tracked the abundance of many different species of woodland caterpillars. He focused on a common predator of caterpillars, the great tit, and recorded how often the birds brought each type of caterpillar home to their nest boxes. He noticed that there was generally a delay of several days between the time when a particular type of caterpillar first hatched in the spring and the time when it began appearing in great tit nests. Tinbergen's explanation was that the birds initially overlooked new prey types because they were not expecting to find them. Their search would be focused on different visual features, and the new prey would be overlooked. When the new prey type became sufficiently abundant, however, some great tits would inevitably come across it by chance, and the birds would switch to a new searching image, focusing on the features of the new prey type. Visual attention would then greatly increase their searching success, and the birds would feed on the new prey exclusively.

This was a remarkable intuitive leap from visual attention to population biology, but Tinbergen's field observations were wide open to other interpretations. In many animals, uncommon food items are eaten less often than you would expect from their actual numbers, while common items are consumed in great excess. This can occur for many reasons, ranging from differences in spatial distribution to predator movement strategies to nutritional constraints. Searching images are a possible interpretation, but they are by no means the only one. Testing behavioral mechanisms inevitably requires more precise experimental controls than one can easily manage in the field. Much of the subsequent attention research in animals has been conducted in controlled

Visual predators, such as the great tit (*Parus major*), find caterpillars efficiently through the use of searching images. Photograph by Hans Menop/Foto Natura (Getty Images).

laboratory experiments, and these later studies strongly supported Tinbergen's intuition. When birds are given a free choice among varying proportions of camouflaged food items, they consistently overlook rare ones and overeat abundant types. The effect occurs only when the food items are hard to detect, which suggests that the birds are using searching images.[3]

At Oxford University in the late 1960s, Marian Stamp Dawkins conducted the first laboratory studies of searching image using domestic chicks. Based on Luuk Tinbergen's conception, Dawkins predicted that if her chicks were using searching images, they would take food items in runs of the same type, starting with one and then switching to another when the first became less common. It would be exactly analogous to the way that people sort hardware. Dawkins dyed rice grains green and orange and sprinkled them over a background of green and orange tiles.

When the chicks fed on the mixed grains, they did not take them in random sequences. They first ate a few of one kind, then a few of the other kind, then they switched back again. Even though the chicks were familiar with both green and orange grains and were able to see both kinds, they still focused their attention on one type at a time. This "runs effect" suggested that searching images could be experimentally induced by presenting visual stimuli one at a time and then manipulating the sequence to produce runs of a particular stimulus type. The idea transformed visual search for multiple prey types into a serial detection task, a psychological procedure in which a subject is presented with a succession of visual displays.

The most compelling data on the use of searching images in animals has come from laboratory work on North American blue jays. Alan Bond, and his colleague, Alan Kamil, raised blue jays in the laboratory and trained them to search for prey items in images projected on computer screens. They presented captive blue jays with various digital moths imbedded in complex textured backgrounds. Only when the jays found and pecked a moth were they rewarded with food.

When the moths were conspicuous, the jays located any type with equal facility and showed no use of searching images. When the moths greatly resembled the background but were very different from one another, the birds focused their attention on one moth type at a time. In this case, they were more likely to detect moths that were similar to ones they had just found. The researchers showed the blue jays as many as several hundred moths that varied in all dimensions, and they placed them on cluttered backgrounds, where the moths were very difficult to find. Under these extreme conditions, the jays relied completely on searching images. Moths that strongly resembled the ones found in recent

This blue jay (*Cyanocitta cristata*) was trained to search for cryptic moths on a computer screen. Photograph by Alan Bond.

trials were detected immediately; moths that strongly contrasted with recent targets were almost invariably overlooked. This study was as close as any lab experiment has come to emulating the difficulties that predators experience in nature. The results suggest that predators commonly rely on visual attention to detect camouflaged items.[4]

Focused searches are essential for finding camouflaged prey, and searching images are automatically generated from the memory of recent successful encounters. But that is really only half the story. Sooner or later, the supply of any particular food item is going to dry up, and continuing to look for yesterday's lunch will be a big disadvantage. What happens when a predator really ought to be looking for something else? How do animals escape from the grip of their searching images? Visual search does not multitask. There is apparently room for only one searching image in short-term memory at a time, and when one image is replaced by another one, the original is lost. But the replacement process seems to get easier over time, since searching images have a limited shelf life. They don't exactly vanish, but they seem to get more porous, more likely to let alternative stimuli leak through. How long a searching image persists may depend on how frequently it has been successful. If you have been finding and eating a caterpillar every five seconds, and then a minute goes by with no new caterpillars, you might be more open to finding other prey. But if you are intensely searching for camouflaged moths and have found one every ten minutes or so, you might tolerate a much longer period of unrewarded searching without getting discouraged.[5]

Leaky visual searches are susceptible to being disrupted by all sorts of conspicuous stimuli. It is another kind of bottom-up guidance, but in this case, it is driven by things that you are *not* looking for. When you have searched too long without results, previously unnoticed items can begin to intrude into the search and draw your attention away from the task at hand. Bright, contrasting colors, unusual shapes, or prey that happen to stand out from the background become more apparent. Natural visual search can thus be seen as a gradual fluctuation among alternative attentional states, with periods of intense focus alternating with times of greater openness to other possibilities. This would help to explain strong searching image effects in blue jays even when the targets were massively and continuously variable in appearance.

The most powerful disruptor of focused search is motion. Not just any motion, but erratic movements that start and stop

Bush katydids (*Scudderia* sp.) can be nearly impossible to see until a single individual moves, then they all become visible. Photograph by Jim McKinley (Getty Images).

abruptly and change direction unpredictably. Such patterns are characteristic of live organisms, and visual systems are exceedingly sensitive to motion from living things. Nocturnal predators are susceptible to even tiny animate changes in their visual field. An owl's visual system readily discounts continuous movement, like the wind in the trees, but the bird's attention will be instantaneously attracted to the slightest twitch of a mouse.[6]

The process of suddenly being able to see camouflaged animals makes a lasting impression. In late summer in Northern California, home gardeners may occasionally hear a brief, high-pitched buzz, something like *bzzzp* coming from the shrubbery. The sound is the courtship call of the male fork-tailed bush katydid. They look like a green grasshopper with long antennae, but without the big angular wings of typical katydids. It is a common species across North America, but it is not often seen: Their call is so brief that its source is frustratingly difficult to locate.

Once, while walking in Berkeley, the authors heard katydid calls coming from a heavily pruned rose bush. It still had a lot of foliage, but the branches were visible from most angles. There were clearly several katydids in the bush because the calls were coming from various parts of the plant, but the insects were invisible. Close inspection revealed nothing but more *bzzzps*. Then, one katydid slightly shifted its hind leg. Abruptly, it became clearly visible. Once the first one was in view, another katydid appeared, and then another and another. There must have been twenty katydids in that one rosebush. It felt like a gray cloud had rolled away, making all of these green insects as obvious as colored balls on a Christmas tree. The leg movement captured visual attention to the location of one insect, and then once a searching image formed, all of the others could be seen. Even knowing beforehand what a katydid looked like did not greatly assist in searching for these highly camouflaged animals. Top-down guidance was insufficient preparation for the visual system. But once attention was captured, perception proliferated throughout the visual field. That is the power of searching images.[7]

When they are broken down into pieces for analysis, visual search and attention seem dauntingly complicated. It is hard to even imagine how the visual system can keep track of everything and make all the necessary adjustments—acquiring searching images, adjusting attentional focus up and down, deciding when to keep attending and when to fade out. But it works, and it works automatically, usually without awareness. When these bottom-up processes happen while you are searching for camouflaged animals, the effect can seem almost magical, like a curtain being pulled back and displaying creatures that only moments before were completely invisible to you.

This helps to address Teddy Roosevelt's second major concern:

What is the advantage of concealing coloration, since camouflaged prey are commonly taken by predators anyway? There is no question that concealment does not provide ironclad protection. Tamarins can learn the cues that will reliably identify cryptic stick insects, focusing on the legs that extend out from their sides. Blue jays can focus their search on the features of camouflaged moths and greatly enhance their ability to detect them. But the cost of focused search is that it limits the range of prey that can readily be found. Attention is like a spotlight: It helps you to see better where the light is shining, but it reduces your ability to see elsewhere in the room. So although predators can find cryptic prey, concealing coloration is still advantageous because a searching image for one type of prey will make alternative species much more difficult, perhaps impossible, to locate. The answer to Roosevelt's challenge is that camouflage helps the survival of a prey animal if the predator is not precisely focused on its critical visual features. This means that at least some of the time the prey escape notice, and when it comes to the evolution of concealing coloration, surviving some of the time is sufficient.

11. Distinctively Different

ALONG NEBRASKA'S PLATTE River, among the hundreds of thousands of sandhill cranes that arrive in early March, you might get lucky and see one or two endangered whooping cranes. The field guide describes how to separate the features of one species from those of the other: The gray plumage and smaller size of the sandhill crane distinguishes it from the white body and black wing tips of the adult whooping crane.[1]

Identifying animals in the wild requires that the observer not be distracted by individual differences but focus instead on the attributes that unite members of the same species. The concept of species is a cherished principle in biology. Species membership is derived from shared genes and common ancestry, and this usually translates into the recognizable features described in field guides. But some animals do not neatly follow the rule that all members of a species should look alike. In some birds, the color differences between the sexes are extreme. Eclectus parrots from New Guinea, Australia, and the South Pacific misled biologists for years because the males are bright green with yellow bills, while females are red and purple with black bills. Until the early twentieth century, they were classified as different species.

During his exploration of the Malay Archipelago in the 1850s, Alfred Russel Wallace was surprised to discover that males and females of the great mormon butterfly were colored so differently that if he had not seen them mating, he would not have known that they were related. The males are identical in appearance across the range of the species; the females, on the other hand, are extraordinarily variable. Some look like the males, others are mimics of other butterfly species, and on one island, each individual female

Some species show extreme color differences between the sexes. **ABOVE:** Eclectus parrot (*Eclectus roratus*) males are bright green with yellow beaks, while females are red with black beaks. Photograph by Tim Laman (Getty Images). **BELOW:** Female great mormons (*Papilio memnon*) mimic other species and are distinctive from the bluish males. Used with permission, Krushnamegh Kunte, PhD, National Center for Biological Sciences (NCBS), Tata Institute of Fundamental Research (TIFR), India.

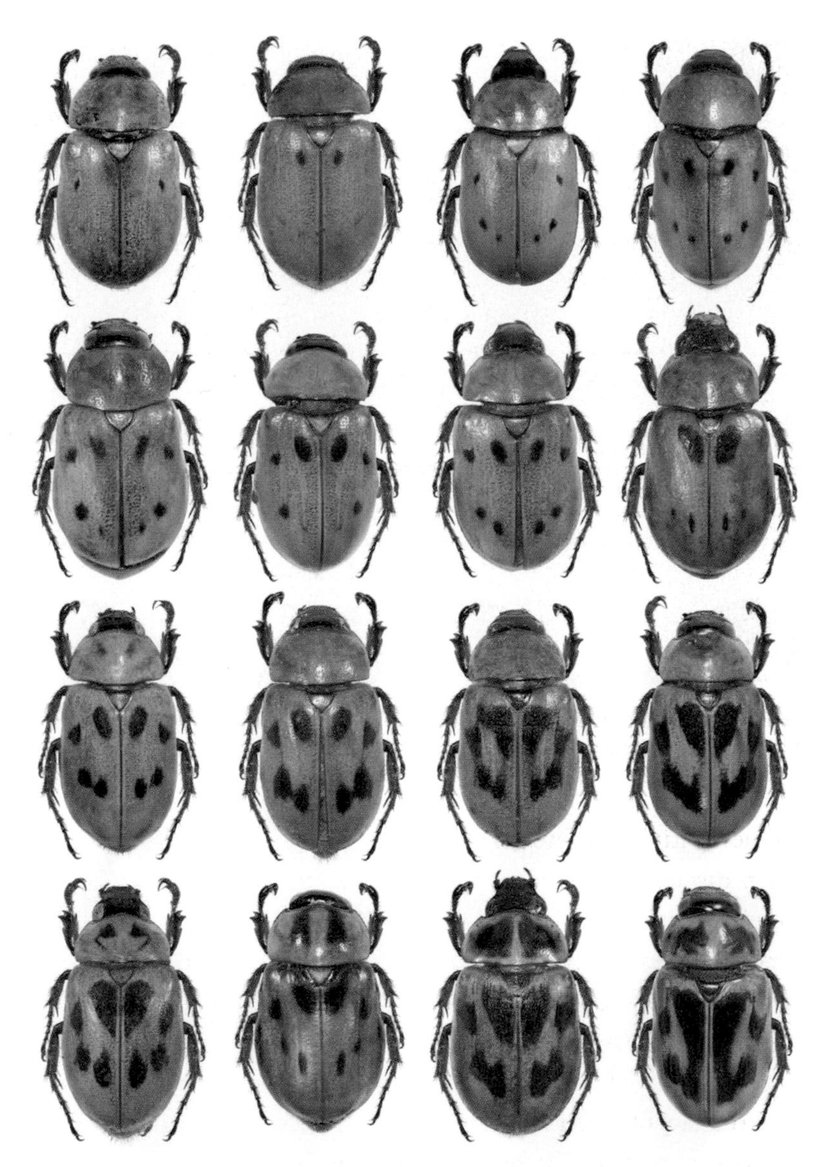

Polymorphic species occur in multiple color variations, as shown left in this scarab (*Cyclocephala sexpunctata*) from Central America. Photograph by Angie Fox, used with permission, Division of Entomology, University of Nebraska State Museum.

Banded snails from a garden in Bulford, England, occur in two different polymorphic species, the white-lipped banded snail (*Cepaea hortensis*) and the banded grove snail (*Cepaea nemoralis*). Photograph by Judy Diamond.

seemed entirely different from every other one. Charles Darwin, when discussing the concept of species, acknowledged, "No one supposes that all the individuals of the same species are cast in the very same mould. These individual differences are highly important for us, as they afford materials for natural selection to accumulate . . ." (Darwin 1859, 45).[2]

Animals that occur in multiple, predictable varieties are referred as polymorphic, which literally means "many morphs" or distinct forms. The term was coined by Kettelwell's mentor, the geneticist E. B. Ford. Although all species show variation in traits, those that are polymorphic elevate this to a fine art. Polymorphism occurs when some forms of a species are selected over many generations to be different from the others. Many factors can drive the evolution of polymorphism. Sex, or more specifically, the selection that occurs as a result of the preferences of one sex, can result in distinct female and male forms, although few are as extreme as eclectus parrots. In Wallace's great mormon butterflies, polymorphism is tied to mimicry. Polymorphism also occurs in a great many camouflaged animals, including species of snails, clams, spiders, crustaceans, and small vertebrates, such as mice, fish, and lizards. Examples can be found in most orders of insects: It has been studied in camouflaged water boatmen, stick insects, beetles, and spittlebugs, and it is pervasive among grasshoppers, mantids, katydids, and moths. More than a third of the species of underwing moths in North America are polymorphic.[3]

Advantages of Deviance

Some camouflaged animals live in a diverse habitat that is a crazy quilt of patches and strongly contrasting colors, and they cannot match all of the patches equally well. Visual predators pick off locally conspicuous individuals, driving selection of more than one color variant. On a uniform background, there is usually an optimal color that conceals animals most effectively from predators. But in a crazy quilt habitat, different color morphs are each protected only on certain patches. The result is a polymorphism of animals that resemble particular background types.

In the California chaparral, Cristina Sandoval, a biologist from the University of California at Santa Barbara, demonstrated how polymorphism protects phasmids that live on shrubs. One color form is green with a white stripe down its back and another is solid green, each matching the color and pattern of a different type of shrub. In Sandoval's experiments, both western scrub jays and western fence lizards consumed more phasmids on nonmatching shrubs. Similar effects have been shown in wrangler grasshoppers, which live in the foothills of the Rockies in northern Colorado. These grasshoppers live on chunks of eroded red granite intermixed among patches of pale green vegetation. They occur in corresponding red and gray-green forms, and each morph is attracted to its own matching substrate. Crazy quilt habitats are also found in the forests that are home to the white underwing moth, whose favorite host plants are birches, aspens, and poplars. The moths occur in five distinctive forms: Three are white with dark markings, a pattern that matches the pale trunks and branches of young trees; the other two are more uniformly dark, resembling the bark color and texture on mature trunks.[4]

Hunting by searching image can also drive the evolution of polymorphism. Predators using searching images are more accurate at detecting prey, and they stay accurate as long as they continue to feed on the same prey form. Their accuracy only declines when they run out of the prey they were seeking and have to switch to something else. As a consequence, predators select prey that are common, and they overlook rare, deviant individuals. Bryan Clarke, a geneticist from the University of Edinburgh, had an important insight: Because of searching image, rare prey morphs are generally ignored by predators, and they will gradually increase in numbers. Only when previously rare prey finally become abundant will the predators begin to take notice. So the abundance of prey morphs will oscillate up and down like a roller coaster as each form takes its turn being the focus of predatory interest. This type of polymorphism is maintained by the relative frequencies of each of the different morphs. It is most common in

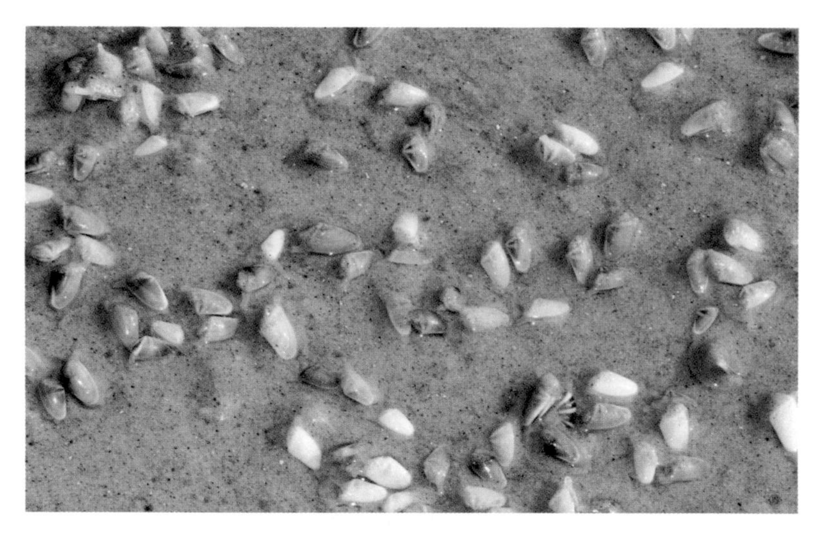

Beach clams (*Donax variabilis*), from North Carolina, show massive polymorphism, where nearly every individual is differently colored. Photograph by Gilbert S. Grant (Getty Images).

species in which the different morphs are equally difficult to find within a single habitat.

Polymorphism tied to searching images can result in a dizzying mixture of morphs, from as few as three to dozens, as in grouse locusts, banded snails, or some underwing moths. The most extreme case is massive polymorphism, where there are so many different forms that no two individuals look alike. The daisy brittle star is broadly distributed from the intertidal to deep water, where it scavenges for food under rocks. At any location, the five fragile arms of individual brittle stars are colored in unique patterns of red, brown, green, and black. Beach clams also occur in immensely variegated colors and patterns. The clams live partly submerged in sand at tide line among gravel and accumulated detritus. Field experiments on carrion crows foraging along European beaches indicate that the birds rely on searching images, suggesting that such predation maintains polymorphism in beach clams.[5]

Carrion crows (*Corvus corone*), on a beach in Scotland, are predators on beach clams and other invertebrates. Photograph by Oxford Scientific (Getty Images).

The Peacock Fish

The northern coast of the island of Trinidad rises abruptly into a range of mountains swathed in rainforest. Numerous rivers and streams cascade down the steep slope to the Caribbean Sea. The most abundant and widely distributed fish in this profusion of fresh water is the guppy. Adult female guppies are about two centimeters in length, with a uniform olive-beige coloration. Males are two-thirds the size of their mates, with color patterns that are a spectacular mosaic of splotches and bars of black, red, orange, or yellow pigment and iridescent areas of blue green or purple. Male color patterns are controlled by a diverse array of genes. Guppy males are massively polymorphic, and no two individuals are exactly alike.

In guppies, male coloration attracts females. To interest a female, a male performs a courtship display, bending his body into an S shape and quivering all over, and then darting off, gradually enticing the female away from the rest of the school. Proper courtship is important for getting results, but one factor that most consistently influences breeding success is the male's color pattern. Females are most responsive to the overall brightness of the males, particularly when they have large patches of red orange. Females are less impressed with small spots and dark coloration. Unfortunately for guppies, they are near the bottom of a food chain that includes predators with excellent vision, such as kingfishers, herons, snakes, and bats. In the small, fast-flowing streams of the northern mountains, they are hunted mainly by cichlids, characins, killifish, and carnivorous freshwater shrimp. Male guppies are caught in an evolutionary quandary: If they are too conspicuous, they are likely to be eaten; if they are too well camouflaged, they will not be able to breed.

The bright colors of the male guppy (*Poecilia reticulata*) attract females. The males are highly polymorphic. Photograph by Dr. Paul V. Loiselle.

The distribution of guppies in Trinidad is patchy and irregular. Some pools have no guppies, others have guppies but no predators, and still others have guppies and multiple predators. The guppy world is divided up into evolutionary islands. Some rivers have fierce predators, and in these, male guppies are found in muted colors. Other rivers have ineffectual predators, and the male guppies are more brightly colored. John Endler, an evolutionary biologist then at Princeton University, saw the potential of this system for investigating color evolution. In a series of classic studies, he established the role of predation and breeding success in determining the color patterns and reproductive behavior of guppies.

Endler began with a field survey. He counted each predatory species in his survey streams and examined their gut contents to determine how often they ate guppies. To determine how readily predators could detect male guppies, Endler measured the spectrum of incident light at various locations and times of day, as well as the light reflected by guppy spots. He then quantified the color pattern of male guppies at each location. If predatory fish concentrate their efforts on conspicuous males, the guppies in rivers with highly effective predators, like cichlids, should have smaller pigmented patches and less iridescence. Guppies in rivers with less efficient predators, like killifish, should be brighter with larger orange spots. Endler's data supported a clear relationship between risk of predation and the coloration of the guppies.

Endler then tested his field observations under controlled laboratory conditions. He set up greenhouse pools floored with aquarium gravel in specific mixtures of colors and two different grit sizes. He introduced populations of guppies that were mixtures of fish from all the rivers he surveyed. Then Endler added predators: either cichlids or killifish or nothing as a control. The results supported what he had observed in the wild. Over the course of a year, the pools with no predators or just killifish showed a shift in the population to more guppies with brighter colored spots. Those in pools with cichlids were less conspicuous and had fewer blue and iridescent patches.. The experimental study also confirmed that guppies' spots are selected to match the size of the gravel background. In the presence of predators, the spot size mirrored the size of the gravel, providing concealment in terms of pattern as well as color.[6]

At first glance, Endler's data suggest that polymorphism in guppies is mainly the result of patchiness in the distribution of predators. But this does not account for the massive polymorphism among guppy males. In 2006, a group of field biologists led by David Reznick from the University of California at Riverside undertook the heroic task of testing whether guppy evolution was also being driven by searching images. The design of the study was similar to Bernard Kettlewell's mark-and-recapture experiment on peppered moths. The scientists established field sites in three small river systems in Trinidad, two with guppies and killifish and one with guppies and cichlids. Within each of their three sites, they located sets of partially isolated pools. They caught all the guppies out of each pool and separated the males from the females. The females went back into the pool they came from. The males were combined across all pools at a site and then divided into two groups, depending on whether or not their tails had a colored spot. Males were returned to pools at their site but in different proportions. So for half the pools, there were three plain-tailed males for every spotted-tailed one; in the other set of pools, the proportions were reversed.

After several weeks, the same laborious process of catching every one of those tiny fish was repeated. The scientists found that the plain-tailed and spotted-tailed males did not differ in overall mortality across sites, but the morph that was initially most common in a given pool was now the least likely to be recaptured. There was a clear, strong bias in the pattern of predation: Predators concentrated their efforts on the locally most common form. These studies suggest that both searching image and environmental patchiness drive the evolution of polymorphism in guppies.[7]

Virtual Ecology

Captive blue jays trained to search for camouflaged digital moths on computer displays show strong evidence of searching images. The effect is most apparent when the jays search on complex, cluttered backgrounds where a wide range of prey morphs are equally difficult to detect. So these controlled laboratory experiments can also test whether hunting by searching image promotes the evolution of polymorphism.

Alan Bond and his colleague, Alan Kamil, created a virtual population of several hundred digital moths of three distinctive morphs and presented them one at a time to the blue jays. If the jay found the moth and pecked it, the bird was given a food reward, and the moth was removed from the population. If the bird overlooked the moth, the moth remained in the population. At the end of each day, surviving moths were replicated to bring the population back to its original level. Over the course of fifty daily generations, the numbers of the three different morphs oscillated smoothly up and down, but none of them was driven to extinction, and no matter what initial proportions were used, the moth morphs ultimately converged on the same relative numbers, showing that the polymorphism was being maintained in the population.

In this laboratory experiment, searching images were able to maintain a polymorphism of digital moths. But could searching images also produce a polymorphism from an initially uniform population? Bond and Kamil created another population of moths, but in this case, each moth was specified by a genome, a string of computer characters that worked like biological genes to code for the appearance of the digital moth. The captive jays were shown moths from an initially uniform population and searched for them as in the earlier study. Surviving moths in this experiment simulated reproduction of real moths, with mutations and recombinations of the two genomes to produce novel offspring. In this experiment, the appearance of the moth population shifted in response to the searching success of the birds. If the jays concentrated on one moth type on one day, those moths would be far less abundant on the next day, and there would be an assortment of novel moths that the birds had never seen before.

Over a hundred generations of selection, the birds produced a moth population that was more camouflaged than the original one but also much more diverse. Blue jays hunting by searching image generated a massive, frequency-based polymorphism, an array of digital moths that were every bit as varied as beach clams or brittle stars. Then the scientists ran a third experiment exposing the virtual moth population to jay predation in a mixed environment of light and dark patches. The effect of the mixed environment was to produce clumped arrays of moths that resembled either the light or dark patches of background. Like male guppies, digital moths show a polymorphism influenced both by searching images and by the patchiness of the environment.[8]

Polymorphism has challenged biologists to explain how multiple forms can evolve within a single species. No single methodology was sufficient to determine the divergent influences of predator search and environmental variability. The explanation turned out to be that both selective factors were at work. Sandoval's field experiments, Clarke's mathematical insights, Endler's and Reznick's field tests and laboratory replications, and Bond's and Kamil's virtual ecology with captive jays together cast light on processes fundamental to the evolution of concealing coloration.

12. Limits to Invisibility

NATURAL SELECTION SHAVES the odds, tipping the balance even slightly in favor of one set of traits over another. Concealing coloration does not need to make an animal immune from predation in order for it to be favored for survival. It just has to make the animal less likely to be detected—on the average and over the long run—than less camouflaged members of the same species. The benefits of concealing coloration are a shift in probability, a slight advantage in the game, so slight that it can be seen only when summed over lifetimes of predator encounters. From the perspective of an individual prey animal, this is pretty cold comfort. In a lava field, a very dark mouse may have a better chance of avoiding detection than a light one, but its odds of living out the year are still very slim. The balance of life and death is not weighed with just an animal's color pattern. What counts in the end are the attributes of the whole organism. When it comes to survival, animals have evolved interconnected suites of features that include color as one component among many.

In 1891, Alfred Russel Wallace wrote *Darwinism,* a summation of his perspective on evolutionary theory over the first thirty years of its history. One chapter was a reflection on animal coloration in which he reasserted his fundamental certainty that so remarkable and conspicuous a trait as color must have arisen under the influence of natural selection and must therefore have some relevance to the welfare of the animals that display it. "But," Wallace regretfully admitted, "the problem is found to be far more complex than was at first supposed" (Wallace 1891, 188). Nearly a century later, John Endler expressed much the same sentiments: "All too often in evolutionary biology we are led to speculate or infer the mode of action of natural selection; we usually do not know why some individuals are more adaptive than others" (Endler 1980, 76). Whatever else one can say about the evolution of concealing coloration, the closer you look at it, the more interesting it becomes.

Concealing coloration is only one tactic in a range of defensive options. Animals remain constantly vigilant, they minimize movement in dangerous circumstances, and they dart into hiding at the slightest disturbance. Theodore Roosevelt repeatedly argued that alertness, stealth, speed, and physical concealment are more important than coloration in ensuring survival. But concealing coloration does not negate such behavioral tactics—it complements them. When concealment fails to work, prey animals resort to other alternatives, ranging from simple avoidance to startling visual displays, irritant chemicals, and aggressive threats. As one tactic fails, the animal switches to another. So dissecting out the effectiveness of concealing coloration alone is like focusing on one particular defensive move in a martial art like taekwondo and asking how it contributes to success in a contest. There are many ways to avoid being eaten, and most species do not depend exclusively on coloration for protection.

Basiceros manni is an odd species of ant from the jungles of Central America. Its entire upper body surface is covered with hairs, shaped rather like whisk brooms and feathers, that capture soil particles and form them into a dry mudpack. *Basiceros* ants become coated with soil, effectively wearing their background. It is an impressively flexible means of concealing coloration, but these insects have evolved an additional protection. Unlike most worker ants, which tend to scurry around at top speed, *Basiceros* workers often stand perfectly still for minutes at a time, not even twitching their antennae. When they do move, it is with the slow,

Basiceros ants maintain concealment by moving extremely slowly, freezing when disturbed, and accumulating fine particles of soil on their backs. Source: Bert Hölldobler and Edward O. Wilson, *The Ants* (Cambridge, Mass.: Belknap Press of Harvard University Press, 1990). Reprinted by permission of Bert Hölldobler.

deliberate steps of a heron stalking a fish. If they are touched even very lightly during their foraging, they freeze, often for several minutes. The pieces of the *Basiceros* survival strategy—the specialized hairs, the slow movements, and the tendency to freeze when disturbed—work together to make these insects nearly undetectable.[1]

A Unified Explanation

The most popular interpretation of animal coloration among early writers was that it was a direct consequence of the action of the environment. Heat and cold, humidity, or dietary ingredients were assumed to transform the colors of animals in the same way that cooking or drying could change the color of food. Pliny the Elder, in his *Historie of the World* in 77 CE, knew that the fur of the arctic hare changed seasonally from brown to white, and he attributed the switch to a change in diet: "Upon the Alpes and such high mountaines, they bee of colour white, so long as the snow lieth; and it is verily thought, that all Winter long they live with eating of snow: for surely, when it is thawed and melted, all the yeare after they be browne and reddish as before" (Plinius Secundus 1601, 232).

Al-Jāḥiẓ, an early Arabic scholar, wrote a compendium of natural history around 776 CE. Like Pliny, he interpreted animal coloration in terms of the direct effects of the environment, excluding the possibility of supernatural influences. In his *Book of Animals,* he emphasized, "We say that God did not make us black ... rather it is our environment that has made us so ... such that the gazelles, ostriches, insects, wolves, foxes, sheep, asses, horses and birds that live there are all black. White and black are the results of environment, the natural properties of water and soil, distance from the sun and intensity of heat" (Al-Jāḥiẓ 1969, 196–97). There was no need to invoke an act of God for something that could so readily have proximal causes.[2]

Environmental explanations can often be contradictory: Desert ravens were thought to be dark because of the heat of the sun, but arctic fur seals were dark because of the coldness of the water. By the eighteenth century, a revolution in scientific thinking had provided unified physical laws that accounted for an array of natural phenomena. Naturalists were seeking similar unified explanations for the diverse elements of the biological world. In *Zoonomia,* Charles Darwin's grandfather, Erasmus, wrote at length on his philosophy of anatomy, physiology, and development. He noted the function of animal color patterns in concealment from predators: "The colours of many animals seem adapted to their purposes of concealing themselves either to avoid danger, or to spring upon their prey" (Darwin 1796, 457).

The underlying mechanism was still out of his reach, however, and Erasmus fell back on an inventive form of individual adaptation. He suggested that all animals might have the color sensitivity of chameleons, at least early in their lives. Somehow the colors perceived by the retina were adopted during development by the organism and were transmitted through "fibers" to the skin, where they were incorporated into hair or feathers. Erasmus admitted, however, that he could not see how his mechanism could possibly work: "The final cause of these colours is easily understood, as they serve some purposes of the animal, but the efficient cause would seem almost beyond conjecture" (Darwin 1796, 458).[3]

Wallace established the modern evolutionary perspective on animal coloration, arguing for a continuity between different types of camouflage and mimicry. He repeatedly argued against any form of direct environmental causation, pointing out numerous situations in which it could not consistently account for animal color patterns. But a few scientists were not convinced by Wallace's argument. The most prominent critic was Frank Beddard, a fellow of the Royal Society and a lecturer at several English universities. Beddard had strong sympathies with Darwinian theory, and he readily accepted object resemblance as a consequence of natural selection, but he was unwilling to accept the universality of natural selection for concealing coloration. In Beddard's view, pigmentation was irrevocably fixed in an animal's biology, so variations in coloration could only be attributed to the direct effects of the environment. The pale golden birds and reptiles of the Sahara must have been brought to a common sandy coloration by the bleaching effects of the sun, and the seasonal change from brown to white in arctic mammals was a clear result of the drop in temperature at the onset of winter.[4]

The relationship between direct environmental effects and natural selection for concealing coloration was finally sorted out in a series of experiments by Edward Poulton, who was the Hope professor of Zoology at Oxford University. Poulton was a determined Darwinist, but he was more measured and cautious than Wallace, and his work placed the field of animal coloration on a more defensible ground. His most important contribution was his research on how environmental influences on coloration operate within the life cycle of animals.

Poulton raised caterpillars of the small tortoiseshell butterfly in the laboratory, and he noticed that if they molted into pupae on black backgrounds, the pupae were dark colored. If they molted on white or shiny backgrounds, the pupae were light. By careful experimentation, Poulton discovered that the choice of which color to adopt was made during the last larval phase, and the color signal was received by the caterpillar's skin, not by its eyes. Even blinded caterpillars would adopt the correct coloration. The association between the animal's color and the background was the result of a physiological process that had evolved to conceal the pupa in differently colored surroundings. He found similar effects in moths, in which the caterpillars developed a color that matched the light reflected from their surroundings. Poulton emphasized that these conditional color strategies could not simply reflect the direct influence of the environment. The ability to respond developmentally to different colors of light had to be inherent in the animal for the environment to exert an influence.[5]

Despite their disagreements, Beddard and Poulton had much in common. They were among the first naturalists to emphasize that organisms were limited by their phylogeny, their genetics, *and* their development. They recognized that animals are not entirely malleable; they cannot evolve an unlimited array of appearances nor respond to any possible selective pressure. And they both understood the importance of control mechanisms that switch between different cryptic colorations. When the environment directly affects coloration, it mainly does so not as a primary physical cause but as a cue that activates a different developmental pathway. The color itself is not inherited, but the mechanism that enables the developmental choice most surely is, and it is this mechanism that has been configured by natural selection. This was the beginning of a major change in evolutionary thinking, the notion that animals could evolve the capability to change their coloration as environmental conditions required. Poulton's butterfly pupae had transcended the primary limitation of concealing coloration, achieving a level of flexibility that would enable them to be protected in a variety of differently colored habitats.[6]

Sometimes the color environment at one stage in development determines the appearance of an animal in another stage. Caterpillars of the small tortoiseshell (*Aglais urticae*) detect the color of their background, and this later determines the color of the pupa. Photograph by Paul Franklin (Getty Images).

The Eighth Plague

In the biblical story of Exodus, the pharaoh refused to release the Israelites and allow them to leave Egypt. His intransigence brought on a series of ten plagues and catastrophes. The eighth one was an invasion of locusts: "[T]hey covered the face of the land, so that the land was darkened, and they ate all the plants in the land and all the fruit of the trees which the hail had left; not a green thing remained, neither tree nor plant of the field, through all the land of Egypt" (*Revised Standard Version,* Exod. 10:15). To farmers in the Middle East, a swarm of desert locusts must truly have seemed a visitation from an angry god. The insects appeared without warning in horrific numbers, descending in whirring clouds on the wheat fields and wreaking massive devastation. But

Nymphs of the desert locust (*Schistocerca gregaria*) develop different color patterns depending on the environmental conditions. When food is abundant, they adopt a pale green "solitary phase" coloration (left). During times of scarcity, the locusts concentrate in large groups, switch to feeding on toxic plants, and adopt a "gregarious phase" coloration (right). Photograph by Gregory Sword.

locust swarms do not appear by magic. Desert locusts are perhaps the single best-understood example of how genetics and development interact with the environment to influence coloration.

Desert locusts live in sparsely distributed areas of brush and bunch grasses in the Sahara Desert, the Arabian Peninsula, and the Sahel region of Africa to the south. In their solitary color phase, they are a uniform pale green with light yellow eyes, resembling the surrounding light green desert scrub. They are relatively inactive and strongly averse to contact with other locusts. When the environment is favorable, locusts can remain in the solitary phase for generations.

But conditions in these dry and unstable habitats never remain constant. When a fluctuation in climate reduces the food supply to smaller clumps, the locusts are forced into closer contact, and contact is the cue that sets off a cascade of changes that will ultimately transform them into entirely different creatures. Even just touching a late-stage locust nymph repeatedly on the hind leg with a fine brush produces a fundamental change in its behavior. Abruptly, the nymph becomes much more active and strongly attracted to other locusts, stimulating them, in turn, to become more social, forming up into groups and marching around together. Sustained contact activates hormonal changes and alters the nervous system. With the decline in the food supply, the locust nymphs begin to eat lower quality plants, many of which have chemicals that are toxic and aversive to vertebrates, and the chemicals add their own influence to the transformation process. At the next molt, the locusts change in appearance as well as behavior.

In their new gregarious color phase, the locusts display a conspicuous black-and-yellow coloration, and they move about aggressively and restlessly in large numbers. They look and behave a bit like wasps, and because they have often eaten toxic plants, their warning coloration may be a valid signal. Although many animals eat locusts in their solitary phase, birds and other predators seldom attack them in their gregarious phase. Under conditions of very low resources and very high population density, gregarious locusts will migrate, and because they are attracted to one another, they migrate as a swarm. At the extreme, this produces millions of locusts flying over hundreds of kilometers and consuming everything in their paths. The economic impact on human populations can be devastating.

The switch between locust phases is truly a transformation. It appears to generate an entirely different life-form, affecting the animal's coloration, morphology, physiology, molecular biology, immunity, and reproduction, and the change is persistent. Locusts that have transformed into the gregarious phase will have gregarious phase offspring, often for several generations, even after the population density has dropped back to normal levels. The gregarious phase appears to have evolved as a density-dependent warning coloration, an antipredator strategy that occurs when camouflage no longer provides adequate protection. It is the ultimate example of flexible defense: Based on behavioral cues, the nymph switches to a different developmental path, discards concealment altogether, and becomes both toxic and warningly colored.[7]

Evolutionary Ghosts

It is not easy to decide whether a particular color pattern is concealing. A photograph or painting of a cryptic animal, even when it is made in an appropriate natural setting, is really only suggestive. The best initial approach is to determine whether the animal's coloration amounts to a random sample of its surroundings, based on the spectrum of incident light and the visual system of the likely predator. Then one has to decide whether it evolved as a consequence of selection by visual predators. Here, the best evidence comes from experiments that test the ability of the predators to detect the animal on various natural backgrounds.

But an animal's coloration does not always reflect its interaction with contemporary predators. Adaptations can persist long after the original selective factors have changed, particularly in long-lived and slow-breeding species. As a result, the color patterns of some animals appear to be residues of interactions with creatures from the distant past. Perhaps the most striking example is the New Zealand kakapo, a large, flightless parrot that is one of the world's most endangered birds. Now confined to isolated island

The kakapo (*Strigops habroptilus*), shown center in its native habitat on Whenua Hou (Codfish) Island, New Zealand, is one of the world's most endangered birds. Its most formidable predator, Haast's eagle (*Harpagornis moorei*), has long been extinct. Photograph by Judy Diamond.

preserves, kakapos were considered extinct as late as 1974, and only recently has the New Zealand Department of Conservation helped to breed back a small population from a few survivors discovered in the wild.

Though they are the size of a small goose, kakapos are surprisingly difficult to see in the wild. Their pale green, mottled plumage makes a perfect match to the leaves, lichen, and mosses of the surrounding forest, and away from their breeding grounds, they are silent. They step slowly and deliberately through the underbrush, and they freeze when disturbed. All these features suggest an evolutionary adaptation to visual predators, but New Zealand today has no native predators that could hunt and kill an adult kakapo. The species was nearly annihilated by mammals, such as feral dogs, cats, stoats, and black rats, introduced by settlers. But the kakapo's coloration far predates European colonization, suggesting that an extinct avian predator may have shaped kakapo evolution. Until about 1350, New Zealand was home to Haast's eagle, the largest raptor ever recorded. The kakapo's mossy coloration might be a kind of fossil camouflage, an adaptation to a predatory interaction that no longer exists.[8]

Evolutionary questions often require multiple, convergent approaches to reveal their dimensions. Where the perceptual abilities of a potential predator are unknown or unknowable, the behavior of the camouflaged animal can help to provide indications of the function of its coloration. Given a choice, does the animal choose to rest on matching backgrounds? Is it found mainly in circumstances in which it is maximally camouflaged? Many animals that show object resemblance, such as sargassum shrimp, orient themselves in a way that enhances their concealment, and highly camouflaged species commonly move more slowly and flush less readily when approached. Finally, the color strategy has to be considered in the context of the life history of the animal. How does it vary with food availability or predation or conflicting selective influences? Concealing coloration is a phenomenon of great richness, and it invites a convergence of disciplines ranging from population ecology and animal behavior to genetics, molecular biology, and biophysics.

The impacts of natural selection on wild populations are well established. Generations of experiments on animal coloration have elevated the insights of Wallace and Darwin to accepted biological fact. But color evolution is only just beginning to be understood. Coloration is influenced by physical and chemical properties, thc behavior and development of the animal, the structure and complexity of the environment, and the visual systems and cognitive abilities of predators and prey. Studying animal coloration is an exercise in time travel, illuminating the conditions of the past that have produced the diversity of the present.

Guide to Common and Scientific Names

COMMON NAMES

agave *Agave* sp.
American badger *Taxidea taxus*
American flamingo *Phoenicopterus ruber*
arctic fox *Vulpes lagopus*
arctic hare *Lepus timidus*
army cutworm *Euxoa auxiliaris*
Asian swallowtail *Papilio xuthus*
bald-faced hornet *Dolichovespula maculata*
banded grove snail *Cepaea nemoralis*
Basiceros ant *Basiceros manni*
beach clam *Donax variabilis, Donacilla cornea*
blue jay *Cyanocitta cristata*
blue morpho *Morpho peleides*
bordered straw *Heliothis peltigera*
bumblebee *Bombus* sp.
cactus finch *Geospiza scandens*
cactus wren *Campylorhynchus brunneicapillus*
California meadow spider *Theridion californicum*
canary *Serinus canaria*
carrion crow *Corvus corone*
cattle, domestic *Bos primigenius*
cerulean sargassum shrimp *Hippolyte coerulescens*
chambered nautilus *Nautilus pompilius*
chameleon Chamaeleonidae
chicken, domestic *Gallus gallus*
chimpanzee *Pan troglodytes*
cicada *Tibicen* sp.

coachwhip *Masticophis flagellum*
collared lizard *Crotaphytus collaris*
common marmoset *Callithrix jacchus*
common starfish *Asterias rubens*
cottontail rabbit *Sylvilagus* sp.
Couch's spadefoot toad *Scaphiopus couchii*
coyote *Canis latrans*
creosote bush *Larrea tridentata*
Cristina's timema *Timema cristinae*
cuttlefish *Sepia* sp.
daisy brittle star *Ophiopholis aculeata*
deer mouse *Peromyscus* sp.
desert lark *Ammomanes deserti*
desert locust *Schistocerca gregaria*
desert mantid *Eremiaphila* sp.
desert sumac *Rhus microphylla*
differential grasshopper *Melanoplus differentialis*
dodo *Raphus cucullatus*
dog, domestic *Canis lupus familiaris*
dolphin fish *Coryphaena hippurus*
Douglas fir *Pseudotsuga menziesii*
earthworm *Lumbricus* sp.
eastern fence lizard *Sceloporus undulatus*
eastern towhee *Pipilo erythrophthalmus*
eclectus parrot *Eclectus roratus*
eland *Taurotragus oryx*
Engelmann spruce *Picea engelmannii*
Eurasian jay *Garrulus glandarius*
fork-tailed bush katydid *Scudderia furcata*
fourwing saltbush *Atriplex canescens*
fox sparrow *Passerella iliaca*

freshwater shrimp *Macrobrachium crenulatum*
fringe-toed lizard *Uma* sp.
frogfish Antennariidae
Galápagos mockingbird *Nesomimus parvulus*
Galápagos tortoise *Chelonoidis nigra*
geometrid moth Geometridae
giant clam *Tridacna maxima*
giant ground sloth *Megalonyx jeffersonii*
giant kelp *Macrocystis* sp.
giraffe *Giraffa camelopardalis*
glass catfish *Kryptopterus bicirrhis*
Glauert's sea-dragon *Phycodurus eques*
goldenrod crab spider *Misumena vatia*
goldfish *Carausius auratus*
gopher snake *Pituophis catenifer*
grasshopper mouse *Onychomys leucogaster*
gray catbird *Dumetella carolinensis*
great mormon *Papilio memnon*
great scallop *Pecten maximus*
great tit *Parus major*
greater roadrunner *Geococcyx californianus*
grizzly bear, brown bear *Ursus arctos*
guppy *Poecilia reticulata*
Haast's eagle *Harpagornis moorei*
hamster *Cricetus cricetus*
hedge sparrow *Prunella modularis*
heliconid butterfly *Heliconius* sp.
honey mesquite *Prosopis glandulosa*
honeybee *Apis* sp.
house wren *Troglodytes aedon*
Indian peafowl *Pavo cristatus*

indigo bunting *Passerina cyanea*
kakapo *Strigops habroptilus*
kangaroo rat *Dipodomys* sp.
katydid Tettigoniidae
kea *Nestor notabilis*
kelp gunnel *Apodichthys sanctaerosae*
lady beetle *Coccinella septempunctata*
lantern bug *Pyrops candelaria*
leaf-mimic katydid *Typophyllum* sp.
leaping guabine *Rivulus hartii*
lesser earless lizard *Holbrookia maculata*
lime hawk-moth *Mimas tiliae*
little bride underwing moth *Catocala micronympha*
little striped whiptail lizard *Aspidoscelis inornata*
loggerhead shrike *Lanius ludovicianus*
longnose leopard lizard *Gambelia wislizenii*
meadow spittlebug *Philaenus spumarius*
meadowlark *Sturnella* sp.
medium ground finch *Geospiza fortis*
millet *Crenicichla alta*
moa Dinornithiformes
mocking-thrush, *see* Galápagos mockingbird
monarch *Danaus plexippus*
monarch flycatcher *Monarcha* sp.
Mormon tea *Ephedra* sp.
mule deer *Odocoileus hemionus*
narrowleaf yucca *Yucca angustissima*
noctuid moth Noctuidae
northern goshawk *Accipiter gentilis*
northern mockingbird *Mimus polyglottos*
northern raven *Corvus corax*

northern rock mouse *Peromyscus nasutus*
nuthatch *Sitta* sp.
obscure grouse locust *Tetrix arenosa*
owl butterfly *Caligo memnon*
parrotfish *Scarus* sp.
peacock mantis shrimp *Odontodactylus scyllarus*
peacock, *see* Indian peafowl
peppered moth *Biston betularia*
phasmid Phasmida
pierid butterfly Pieridae
pinkedged sulphur *Colias* sp.
pinyon pine *Pinus edulis*
plains pocket mouse *Perognathus flavescens*
plains zebra *Equus quagga*
ponderosa pine *Pinus ponderosa*
ptarmigan *Lagopus* sp.
purple-crested turaco *Tauraco porphyreolophus*
pygmy mammoth *Mammuthus exilis*
pygmy seahorse *Hippocampus bargibanti*
red-crested tamarin *Saguinus geoffroyi*
red-spotted newt *Notophthalmus viridescens*
red-tailed hawk *Buteo jamaicensis*
ring-tailed cat *Bassariscus astutus*
robber fly *Holcocephala fusca*
robin *Erithacus rubecula*
rock pocket mouse *Chaetodipus intermedius*
rock squirrel *Spermophilus variegatus*
Rocky Mountain bristlecone pine *Pinus aristata*
rose-breasted grosbeak *Pheucticus ludovicianus*
sand-burrowing beetle *Chaerodes trachyscelides*
sandgrouse *Pterocles* sp.

sandhill crane *Grus canadensis*
sargassum *Sargassum natans, S. fluitans*
sargassum fish *Histrio histrio*
scarab *Cyclocephala sexpunctata*
scarlet tiger *Callimorpha dominula*
short-horned grasshopper Acrididae
short-horned lizard *Phrynosoma hernandesi*
side-blotched lizard *Uta stansburiana*
silver birch *Betula pendula*
slender sargassum shrimp *Latreutes fucorum*
small tortoiseshell *Aglais urticae*
smooth bromegrass *Bromus inermis*
snowberry clearwing *Hemaris diffinis*
soaptree yucca *Yucca elata*
song thrush *Turdus philomelos*
southern elephant seal *Mirounga leonina*
southern plains woodrat *Neotoma micropus*
spiky grouse locust *Discotettix* sp.
spineless horsebrush *Tetradymia canescens*
spotted flycatcher *Muscicapa striata*
spotted ground squirrel *Spermophilus spilosoma*
spotted hyena *Crocuta crocuta*
starling *Sturnus vulgaris*
stick caterpillar *Ennomos alniaria, E. quercinaria*
stoat *Mustela erminea*
Swainson's hawk *Buteo swainsoni*
tiger beetle Cicindelinae
tiger-striped longwing *Heliconius ismenius clarescens*
tree lizard *Urosaurus ornatus*
twospot astyanax *Astyanax bimaculatus*
tyrant flycatcher Tyrannidae

Utah juniper *Juniperus osteosperma*
Virginia opossum *Didelphis virginiana*
water boatman Corixidae
water strider *Gerris sp.*
western fence lizard *Sceloporus occidentalis*
western rattlesnake *Crotalus viridis*
western scrub jay *Aphelocoma californica*
White Sands camel cricket *Ammobaenites arenicolus*
white underwing *Catocala relicta*
white-crowned sparrow *Zonotrichia leucophrys*
white-lipped banded snail *Cepaea hortensis*
white-throated sparrow *Zonotrichia albicollis*
white-throated woodrat *Neotoma albigula*
whooping crane *Grus americana*
willet *Catoptrophorus semipalmatus*
wood pigeon *Columba palumbus*
wrangler grasshopper *Circotettix rabula*
yellowhammer *Emberiza citrinella*
yellowtail amberjack *Seriola lalandi*

SCIENTIFIC NAMES

Accipiter gentilis northern goshawk
Acrididae short-horned grasshoppers
Agave sp. agave
Aglais urticae small tortoiseshell
Ammobaenites arenicolus White Sands camel cricket
Ammomanes deserti desert lark
Antennariidae frogfish
Aphelocoma californica western scrub jay
Apis sp. honeybee
Apodichthys sanctaerosae kelp gunnel

Aspidoscelis inornata little striped whiptail lizard
Asterias rubens common starfish
Astyanax bimaculatus twospot astyanax
Atriplex canescens fourwing saltbush
Basiceros manni Basiceros ant
Bassariscus astutus ring-tailed cat
Betula pendula silver birch
Biston betularia peppered moth
Bombus sp. bumblebee
Bos primigenius domestic cattle
Bromus inermis smooth bromegrass
Buteo jamaicensis red-tailed hawk
Buteo swainsoni Swainson's hawk
Caligo memnon owl butterfly
Callimorpha dominula scarlet tiger
Callithrix jacchus common marmoset
Campylorhynchus brunneicapillus cactus wren
Canis latrans coyote
Canis lupus familiaris domestic dog
Carausius auratus goldfish
Catocala micronympha little bride underwing
Catocala relicta white underwing
Catoptrophus semipalmatus willet
Cepaea hortensis white-lipped banded snail
Cepaea nemoralis banded grove snail
Chaerodes trachyscelides sand-burrowing beetle
Chaetodipus intermedius rock pocket mouse
Chamaeleonidae chameleons
Chelonoidis nigra Galápagos tortoise
Cicindelinae tiger beetles
Circotettix rabula wrangler grasshopper

Coccinella septempunctata lady beetle
Colias sp. pinkedged sulphur
Columba palumbus wood pigeon
Corixidae water boatmen
Corvus corax northern raven
Corvus corone carrion crow
Coryphaena hippurus dolphin fish
Crenicichla alta millet
Cricetus cricetus hamster
Crocuta crocuta spotted hyena
Crotalus viridis western rattlesnake
Crotaphytus collaris collared lizard
Cyanocitta cristata blue jay
Cyclocephala sexpunctata scarab
Danaus plexippus monarch
Didelphis virginiana Virginia opossum
Dinornithiformes moas
Dipodomys sp. kangaroo rat
Discotettix sp. spiky grouse locust
Dolichovespula maculata bald-faced hornet
Donacilla cornea beach clam
Donax variabilis beach clam
Dumetella carolinensis gray catbird
Eclectus roratus eclectus parrot
Emberiza citrinella yellowhammer
Ennomos alniaria stick caterpillar
Ennomos quercinaria stick caterpillar
Ephedra sp. Mormon tea
Equus quagga plains zebra
Eremiaphila sp. desert mantid
Erithacus rubecula robin

Euxoa auxiliaris army cutworm
Gallus gallus domestic chicken
Gambelia wislizenii longnose leopard lizard
Garrulus glandarius Eurasian jay
Geococcyx californianus greater roadrunner
Geometridae geometrid moths
Geospiza fortis medium ground finch
Geospiza scandens cactus finch
Gerris sp. water strider
Giraffa camelopardalis giraffe
Grus americana whooping crane
Grus canadensis sandhill crane
Harpagornis moorei Haast's eagle
Heliconius ismenius clarescens tiger-striped longwing
Heliconius sp. heliconid butterfly
Heliothis peltigera bordered straw
Hemaris diffinis snowberry clearwing
Hippocampus bargibanti pygmy seahorse
Hippolyte coerulescens cerulean sargassum shrimp
Histrio histrio sargassum fish
Holbrookia maculata lesser earless lizard
Holcocephala fusca robber fly
Juniperus osteosperma Utah juniper
Kryptopterus bicirrhis glass catfish
Lagopus sp. ptarmigan
Lanius ludovicianus loggerhead shrike
Larrea tridentata creosote bush
Latreutes fucorum slender sargassum shrimp
Lepus timidus arctic hare
Lumbricus sp. earthworm
Macrobrachium crenulatum freshwater shrimp

Macrocystis sp. giant kelp
Mammuthus exilis pygmy mammoth
Masticophis flagellum coachwhip
Megalonyx jeffersonii giant ground sloth
Melanoplus differentialis differential grasshopper
Mimas tiliae lime hawk-moth
Mimus polyglottos northern mockingbird
Mirounga leonina southern elephant seal
Misumena vatia goldenrod crab spider
Monarcha sp. monarch flycatcher
Morpho peleides blue morpho
Muscicapa striata spotted flycatcher
Mustela erminea stoat
Nautilus pompilius chambered nautilus
Neotoma albigula white-throated woodrat
Neotoma micropus southern plains woodrat
Nesomimus parvulus Galápagos mockingbird
Nestor notabilis kea
Noctuidae noctuid moths
Notophthalmus viridescens red-spotted newt
Odocoileus hemionus mule deer
Odontodactylus scyllarus peacock mantis shrimp
Onychomys leucogaster grasshopper mouse
Ophiopholis aculeata daisy brittle star
Pan troglodytes chimpanzee
Papilio memnon great mormon
Parus major great tit
Passerella iliaca fox sparrow
Passerina cyanea indigo bunting
Pavo cristatus Indian peafowl (peacock)
Pecten maximus great scallop

Perognathus flavescens plains pocket mouse
Peromyscus nasutus northern rock mouse
Peromyscus sp. deer mouse
Phasmida phasmids
Pheucticus ludovicianus rose-breasted grosbeak
Philaenus spumarius meadow spittlebug
Phoenicopterus ruber American flamingo
Phrynosoma hernandesi short-horned lizard
Phycodurus eques Glauert's sea-dragon
Picea engelmannii Engelmann spruce
Pieridae pierid butterflies
Pieris rapae cabbage white butterfly
Pinus aristata Rocky Mountain bristlecone pine
Pinus edulis pinyon pine
Pinus ponderosa ponderosa pine
Pipilo erythrophthalmus eastern towhee
Pituophis catenifer gopher snake
Poecilia reticulata guppy
Prosopis glandulosa honey mesquite
Prunella modularis hedge sparrow
Pseudotsuga menziesii Douglas fir
Pterocles sp. sandgrouse
Pyrops candelaria lantern bug
Raphus cucullatus dodo
Rhus microphylla desert sumac
Rivulus hartii giant rivulus
Saguinus geoffroyi red-crested tamarin
Sargassum fluitans sargassum
Sargassum natans sargassum
Scaphiopus couchii Couch's spadefoot toad
Scarus sp. parrotfish

Sceloporus undulatus eastern fence lizard
Sceloporus occidentalis western fence lizard
Schistocerca gregaria desert locust
Scudderia furcata fork-tailed bush katydid
Sepia sp. cuttlefish
Serinus canaria canary
Seriola lalandi yellowtail amberjack
Sitta sp. nuthatch
Spermophilus spilosoma spotted ground squirrel
Spermophilus variegatus rock squirrel
Strigops habroptilus kakapo
Sturnella sp. meadowlark
Sturnus vulgaris starling
Sylvilagus sp. cottontail
Tauraco porphyreolophus purple-crested turaco
Taurotragus oryx eland
Taxidea taxus American badger
Tetradymia canescens spineless horsebrush
Tetrix arenosa obscure grouse locust
Tettigoniidae katydids
Theridion californicum California meadow spider
Tibicen sp. cicada
Timema cristinae Cristina's timema
Tridacna maxima giant clam
Troglodytes aedon house wren
Turdus philomelos song thrush
Typophyllum sp. leaf-mimic katydid
Tyrannidae tyrant flycatchers
Uma sp. fringe-toed lizard
Urosaurus ornatus tree lizard
Ursus arctos grizzly bear, brown bear

Uta stansburiana side-blotched lizard
Vulpes lagopus arctic fox
Yucca angustissima narrowleaf yucca
Yucca elata soaptree yucca
Zonotrichia albicollis white-throated sparrow
Zonotrichia leucophrys white-crowned sparrow

Notes

1. Disappearing Acts

1. General reviews of concealing coloration are in Cott (1957), Portman (1959), Owen (1980), Edmunds (1974, 1990), Ruxton, Sherrat, and Speed (2004), Caro (2005), Naskrecki (2005), Forbes (2009), Théry and Gomez (2010), and Stevens and Merilaita (2011). Physiological functions of coloration are from Hamilton (1973), Needham (1974), Prota (1992), and Bortolotti (2006). The literature on the use of color in social communication and advertisement has exploded over the past decade (reviews in Rosenthal 2007, Bradbury and Vehrencamp 2011). Evolutionary issues involved in color signaling are discussed in Endler (1993), Searcy and Nowicki (2005), and Endler et al. (2005). Ruxton (2011) discusses the effects of camouflage involving senses other than vision. Olfactory mimicry is in Czaplicki, Porter, and Wilcoxon (1975) and Dobkin (1979), pheromonal mimicry in snakes is described by Mason and Crews (1986), and vocal mimicry is discussed in Payne (1973, 1976) and Payne, Payne, and Woods (1998).
2. Tristram's biography is from Baker (1996), his travel narrative is in Tristram (1860), and his discussion of the evolution of desert birds is in Tristram (1859). Tristram's account of his bird research, "On the Ornithology of Northern Africa," was spread across four of the first issues of *Ibis,* the house organ of the British Ornithologists' Union, of which Tristram was one of the founders (Buckland 2004).
3. The article Tristram read was Darwin and Wallace (1858). The first edition of *The Origin of Species* was Darwin (1859).
4. In the first edition of the *Origin,* Darwin only briefly discussed the evolution of protective coloration, illustrated by different grouse species that generally resemble the color of their surroundings (Darwin 1859, 84–85). Darwin's November 10, 1862, letter to Bates indicates how profoundly his ideas had been changed (Darwin 1862). The Amazonian expedition is described in Bates (1864) and Raby (2001). Bates's account of butterfly mimicry is detailed in Bates (1862). The Pieridae includes familiar species like the cabbage white and sulphur butterflies. Heliconids are mainly confined to the New World tropics, but they are common in butterfly exhibits at zoos and science centers, and several species are widely distributed in the southeastern United States and southern Texas. Since the 1800s, Batesian

mimicry has grown to an enormous and dynamic field of research (Wickler 1968). For information about mimicry in *Heliconia* in particular, see Mallet and Joron (1999), Joron et al. (2006), Kronforst and Gilbert (2008), and Hines et al. (2011). On mimicry in vertebrates, see Greene (1977), Greene and McDiarmid (1981), Pough (1988), Cheney and Marshall (2009), and Kikuchi and Pfenning (2010).

5. See Wallace (1858). The incident is described, from differing perspectives, in Desmond and Moore (1991) and Raby (2001).
6. Wallace's argument for the continuity between mimicry and other forms of protective coloration was originally proposed in his 1867 article in the *Westminster Review* (Wallace 1867). See also Wallace (1866).
7. By the sixth edition of the *Origin,* Darwin had come to share Wallace's color enthusiasms: "... Insects are preyed on by birds and other enemies, whose sight is probably sharper than ours, and every grade in resemblance which aided an insect to escape notice or detection, would tend towards its preservation; and the more perfect the resemblance so much the better for the insect" (Darwin 1872, 182). In the first edition of the *Origin* in 1859, the only reference to *evolution* was the last word of the last sentence in the book: [F]rom so simple a beginning endless forms most beautiful and most wonderful have been, and are being, evolved" (Darwin 1859, 490). There were no changes in the frequency of this term through the fourth edition and only one additional usage in the fifth. But by the sixth edition, in 1872, there were six uses of *evolve* or *evolved* and ten of *evolution* or *evolutionist.* The term had clearly become established.
8. Brinkley (2009) describes Roosevelt's youth and scientific career. Roosevelt's review is from Roosevelt (1910, 1911b). The remark on Roosevelt's presidential publication record was from Gould (1991). Roosevelt was not unique in his mixture of science and politics. In 1799, during his presidential campaign, Thomas Jefferson published an interpretation of the newly discovered fossils of the giant ground sloth in the *Transactions of the American Philosophical Society* (Greene 1959). The depth of Roosevelt's commitment to nature was quite extraordinary. In "My Life as a Naturalist," Roosevelt reminisced, "I can no more explain why I like 'natural history' than why I like California canned peaches.... All I can say is that almost as soon as I began to read at all I began to like to read about the natural history of beasts and birds" (Roosevelt 1918b, 321–22).
9. Roosevelt originally outlined his ideas about animal coloration in an appendix to the narrative of his African safari (Roosevelt 1910), as a response to the recent publication of Gerald and Abbott Thayer's *Concealing-Coloration*

in the Animal Kingdom (Thayer 1909). It was, at least for Roosevelt, an unusually kind and conciliatory discussion. But when Thayer responded with contempt, the gloves came off. Roosevelt greatly amplified both the argument and his vitriol in his *American Museum* monograph (Roosevelt 1911b), and he revisited the subject repeatedly in print (Roosevelt and Heller 1914; Roosevelt 1918a, 1918b). Many of his most strongly worded claims refer to African big game species, with which, as a hunter, he was exceedingly well acquainted (Roosevelt 1918a, 1918b). Roosevelt's primary misgivings are described in a letter to Charles Kofoid (Roosevelt 1911a). An account of Roosevelt and Thayer's substantial disagreements about animal coloration is described in Gould (1991).

10. Judd's museum study is from Judd (1899). Judd spent his career in the Division of Economic Ornithology, part of the U.S. Biological Survey, working under C. Hart Merriam (Judd 1905).

2. Mistaken Identity

1. Rachel Carson provided a wonderful description of the Sargasso Sea in her landmark book, *The Sea Around Us* (1951). It is an area of the North Atlantic bounded by three great ocean currents—the Gulf Stream on the west, along the East Coast of North America; the Equatorial Current on the south, at roughly the latitude of Puerto Rico; and the return current from the Azores and Canaries on the east (Dixon 1925). The *Sargassum* genus of brown macroalgae is fairly diverse, containing species that attach to the ocean floor in shallow water or on coral reefs, as well as those that never attach to the seafloor during their life cycle (Coston-Clements et al. 1991). More information about Glauert's sea-dragons (figure 2.1) is in Connolly et al. (2002).
2. Peter Osbeck returned to Sweden with about 600 specimens for the Linnaean Herbarium, including 26 species that were described for the first time. Linnaeus later used many of Osbeck's names and descriptions in his *Species Plantarum* (Merrill 1916). Osbeck's expedition diary, entitled *A Voyage to China and the East Indies* (1771), was filled with systematic accounts of the plants he encountered, providing the first scientific documentation of many of the wild and cultivated plants of China and Eastern Asia.
3. The biology of the sargassum fish is more complicated, and more fascinating, than Osbeck imagined (Dixon 1925; Adams 1960; Randall 2005).

Distribution and feeding techniques of frogfish are described in Pietsch and Grobecker (1987). The sargassum nursery is essential to many open-water species (Coston-Clements et al. 1991).

4. Alexander Emanuel Agassiz was the son of Louis Agassiz, who is credited with being the first scientist to propose the existence of ice ages. Louis was the director of the Museum of Comparative Zoology at Harvard from its founding in 1859 until his death in 1873 (Williams 1991; Winsor 1991). Alexander was named a curator at the museum a year after his father's death. Agassiz (1888) describes his 1877 biological survey of the Sargasso Sea.
5. The shrimp experiment is described in full in Hacker and Madin (1991).
6. Object resemblance has undergone its own terminological evolution. Kirby and Spence (1843) referred to mimicry when describing the appearance of stick insects. This was the first use of the term in the scientific literature, and it is still common today to refer to stick mimics or leaf mimics. Wallace's colleague Edward Poulton (1890) called it special resemblance, but Wallace (1867) had used that term to refer to Batesian mimicry, which is a very different phenomenon. The Cambridge University zoologist Hugh Cott, in his *Adaptive Coloration in Animals* (Cott 1957), cites a huge array of examples of object resemblance, including leaf mimicry. Leaf insects are described in Otte and Brock (2003), tropical katydids in Castner and Nickle (1995), and leaf-mimic katydids in Naskrecki (2005). Robinson (1979) discusses the role of predation in promoting prey diversity in tropical habitats.
7. Wallace on behavior as an element of object resemblance is from Wallace (1867). Development of sargassum shrimp is from Hacker and Madin (1991).

3. Pepper and Smoke

1. The European infatuation with natural history is vividly described in Barber (1980) and Thomas (1983). The oldest entomological society in England was the Aurelian Club, established in the 1690s (Harris 1766; Allen 1976). Amateur collectors have made significant contributions to natural history museum collections, particularly in entomology and paleontology (Diamond and Kociolek 2012). In the 1950s, amateur collecting was sometimes ridiculed by scientists (Johnson 2007). Ernst Mayr commented that, "There have always been two kinds of taxonomists: those for whom species of animals and plants are like postage stamps to be collected or described

merely for the sake of rarity or novelty, and those others for whom systematics is an important branch of biology" (Mayr 1955, 47).

2. During development, peppered moth caterpillars flexibly adopt the general coloration of their host plant. The change is apparently mediated by visual experience, rather than diet, and it can be reversed each time the caterpillar molts (Noor, Parnell, and Grant 2008).
3. The story of the peppered moth (*Biston betularia*) is one of the classic biological narratives. It is told in more detail in Kettlewell (1973), Majerus (1998, 2002), Grant (1999), and Cook (2003). Peppered moths were originally described by Linnaeus (1758, 521). All four specimens in the Linnean type collection are typical, light colored forms (www.linnean-online.org; accessed February 7, 2012). An engraver and entomologist, Moses Harris, described the peppered moth in 1766. Although Harris mentions that the caterpillar is found in varying colors, he does not write of color variation in the adults, suggesting that if dark forms were present, they may have been so rare that they had not been noticed (Harris 1766). The melanic form of the moth was fully described and illustrated by Pierre Millière in his *Iconographie et description de Chenilles et Lépidoptères inédits* (Millière 1864, 228), though priority is often given to Edleston (1865) based on a brief letter from 1864. Millière described a dark specimen collected in northern England by Jules Ferdinand Fallou, a prolific amateur with one of the largest personal collections of Lepidoptera in Europe. Edleston (1865) remarked that some sixteen years before, the dark form was almost unknown around Manchester, but more recently it had become common, and by the time of his writing, it was pervasive. When the melanic morph first occurred is a matter of conjecture. The earliest peppered moth collections at the Oxford University Museum of Natural History, dating from before 1818, have many light specimens, but there are also a few darker and medium brown ones, suggesting that this species might have exhibited some color variation from very early times. A vivid account of the magnitude of smoke pollution in nineteenth-century England is in Mosley (2008). The recent decline in numbers of dark moths has been detailed in Brakefield (1990a), Grant, Owen, and Clark (1996), Grant et al. (1998), Cook, Dennis, and Mani (1999), and Saccheri et al. (2008).
4. Tutt's discussion of peppered moth coloration is in Tutt (1896). The alternative explanations are discussed in Kettlewell (1973), Brakefield (1987), Berry (1990), and Majerus (1998). Regardless of their strategies for concealment, moths as a group are savored by a wide range of predators who pick them out of their hiding places in the daytime or snag them in the air when the moths

are in motion (Majerus 2002). Moths are taken by a wide array of birds, lizards, and even grizzly bears, which feed on vast quantities of hibernating adults of the army cutworm (Mace and Jonkel 1986). At night, moths are captured in flight by bats, owls, and nighthawks (Brigham 1990).

5. Kettlewell studied in Cambridge from 1926 and was an active member of the Cambridge University Natural History Society (Wiltshire 1979, 101). In the 1930s, he published articles on the natural history and breeding of less notorious British Lepidoptera, such as the bordered straw (*Heliothis peltigera*) and the scarlet tiger (*Callimorpha dominula*; Wiltshire 1979). E. B. Ford had collaborated extensively with R. A. Fisher in developing his ideas about ecological genetics, and Ford's book, *Mendelism and Evolution* (Ford 1953), was the state of the art at the time Kettlewell was hired. Ford's offer to Kettlewell is in Ford (1948), cited with permission of the president and Fellows of Wolfson College, Oxford. After Kettlewell returned to Oxford, he reestablished his connections to British entomology clubs, eventually calling on amateur entomologists throughout Britain to conduct national surveys of melanic frequencies in moth populations (Kettelwell 1973; Lees 1979; Grant 1999). When Kettlewell died after a long illness in 1979, Ford called him, "The finest living ecologist of the Lepidoptera" (Ford 1979, 166).
6. Peppered moth coloration is now known to be controlled by a single regulatory gene (van't Hof and Saccheri 2010) with four mutant alleles, each of which produces progressively greater melanic effects (Lees and Creed 1977; Cook and Grant 2000; Grant 2004). Kettlewell's ecological surveys are summarized in Kettlewell (1958, 1973) and Kettlewell and Conn (1977). More recent work has confirmed a strong correlation between the numbers of dark moths and the local concentration of sulfur dioxide (reviewed in Cook 2003). Kettlewell's studies are in Kettlewell (1955, 1956). Deanend is an old woodland of huge beech and oak trees planted during the reign of George III. Far from the smoke-belching factories of Birmingham, these Dorset woods contained pollution-free trees still covered with light-colored lichen. Kettlewell's (1955, 1956) mark and recapture experiments in contrasting environments were an elegant innovation, generating compelling data. In polluted Birmingham, 123 of the 447 released dark moths were recaptured (28 percent), but only 18 of the 137 pale moths (13 percent). In unpolluted Dorset, the reverse was true: 54 of the 393 pale moths were recaptured (14 percent), but only 19 of the 406 dark ones (5 percent). Both differences were significant and both were in the direction predicted by Kettlewell's hypothesis (Kettlewell 1955, 1956).

7. The mark and recapture studies have been replicated in at least six subsequent experiments in various parts of England. Cook (2000) reviewed all previous estimates of mortality from these studies and found a reliable correspondence across areas between the local frequency of a morph and the probability of its survival, supporting Kettlewell's inference. As Ridley (2004, 113) notes, "The fitness estimates for the two forms of the peppered moth are about the most repeated result in evolutionary biology." The concern about where peppered moths tend to settle in nature seems to have been resolved (Liebert and Brakefield 1987; Grant 1999; Majerus 2005, 2009), though the possible effects of many other factors remain open issues. Arguing over potential experimental confounds is standard operating procedure in science, but it can be confusing for nonscientists. A variety of popular accounts of Kettlewell's work have played on public misunderstanding, using inaccurate and inflammatory comments to promote unjustified perceptions of incompetence, fraud, and deception (Coyne 1998; Wells 1999; Hooper 2002). But the legitimate scientific arguments in Kettlewell's case were all about minor details. No authoritative researcher has ever seriously disputed his basic findings (Grant 1999; Cook 2000; Majerus 2009), and the accusations of fraud have been roundly refuted (e.g. Grant 1999; Rudge 2005; Majerus 2009).
8. Majerus's study is described in Majerus (2007, 2009) and in Cook et al. (2012). Majerus died unexpectedly before being able to complete his last seasons of data acquisition. There are still many details to be explored in the peppered moth story. Crypticity is as much a result of confusion and clutter as of background similarity (Merilaita 2003; Bond 2007), so dark moths may be fairly inconspicuous even in unpolluted forest. This suggests that selection may not be equally strong in both directions (Clarke and Sheppard 1966). Pollution could also select for melanic coloration through a number of nonvisual effects on fitness (Brakefield 1987; Berry 1990; Lees, Creed, and Duckett 1973; Steward 1977), providing physiological benefits in soot-enshrouded forests. In a series of population models, reviewed in Mani (1990), Mani integrated historical abundance data with measured levels of sulfur dioxide and parameters from experimental studies. The evolutionary dynamics could only be accounted for by a balance of migration and both visual and nonvisual selection (Mani and Majerus 1993). The long-distance migrations of adult and larval peppered moths are discussed in Brakefield (1987), Brakefield and Liebert (1990), Majerus (1998), and Saccheri et al. (2008).

4. Obscured by Patterns

1. Gertrude Stein recalls her conversation with Picasso: "I very well remember at the beginning of the war being with Picasso on the boulevard Raspail when the first camouflaged truck passed. It was at night, we had heard of camouflage but we had not yet seen it and Picasso amazed looked at it and then cried out, yes it is we who made it, that is cubism." (Stein 1984, 11; see also Blechman 2004, 276). Stein went on to imply that the entire First World War had been something of an exercise in cubism (Wright 2005).
2. The history of military camouflage is discussed in Behrens (2002), Blechman (2004), and Forbes (2009).
3. The life and contributions of Abbott H. Thayer are reviewed in Boynton (1952) and Nemerov (1997). Thayer was strongly opposed to the modernist art of the early twentieth century, calling it obscene, sensual, and decadent. White (1951) quotes a pungent letter from Thayer to the editor of *Bruno's Weekly*, roundly condemning an Aubrey Beardsley painting on the cover of the magazine. Thayer's ideas on concealing coloration were presented in published articles and in the book, *Concealing-Coloration in the Animal Kingdom*, written by his son, Gerald (Thayer 1909). There are numerous examples of Thayer's letters to Franklin Roosevelt (the assistant secretary of the U.S. Navy) and Arthur Balfour (the first lord of the British Admiralty) in his collected correspondence in the Smithsonian Institution's Archives of American Art. One of Thayer's last books was *Exhibition of Pictures Illustrating Protective Coloration in Nature and Concerned with the Origination of Camouflage in War* (Thayer and Thayer 1919), which summarizes his contribution to the development of military camouflage. Thayer's philosophy of art and reality is described sympathetically by Cortissoz (1923) and Fuertes (1910) and more critically discussed in Kingsland (1978), Nemerov (1997), and Gephart (2001). Much of the popular acceptance of disruptive coloration was derived from Cott's (1957) massive compendium of coloration strategies.
4. The Stevens and Cuthill experiments on disruptive coloration are described in Cuthill et al. (2005, 2006), Stevens and Cuthill (2006), and Stevens (2007). Later studies of other types of visual disruption are in Cuthill and Székely (2009) and Stevens et al. (2009). See also Osorio and Srinivasan (1991) and Stevens and Merilaita (2011).
5. Thayer's discovery was originally published in Thayer (1896a, 1896b). His presentation at the AOU meeting in 1896 is described by the society's secretary (Sage 1897, 85–86). After his trip to England, Thayer published a

revised version of his 1896 paper, which differs strikingly from the original in its repeated invocation of evolutionary principles (Thayer 1902). Thayer's insistence on using imagined artistic renderings as if they were legitimate scientific evidence ultimately led to criticism. His most articulate and vigorous opponent was Theodore Roosevelt (particularly 1911b). The rigid, condescending artist and the volatile, combative politician grated on each other, and they spent years exchanging angry diatribes (Gould 1991). What infuriated Roosevelt was Thayer's use of illustrations that were made under conditions that were decidedly abnormal: "[M]uch of Mr. Thayer's work is predicated on experiments which stand on a par with putting a raven or a magpie into a coal scuttle in order to show that its coloration is concealing" (Roosevelt 1911b, 132).

6. The problems that plague any universal evolutionary interpretation of countershading are reviewed in Kiltie (1988) and Ruxton, Speed, and Kelly (2004). De Ruiter's experiments on caterpillars are in de Ruiter (1952, 1955). Rowland's experiments are described in Rowland et al. (2007, 2008) and Rowland (2009, 2011). The original experiments on countershaded caterpillars made of pastry were by Edmunds and Dewhirst (1994). See also Young and Roper (1976).
7. The relative influence of color and pattern on crypticity is reviewed in Bond and Kamil (2006) and Bond (2007). Endler (1978, 1984) defined an animal's coloration as cryptic if it resembles a random sample of the background, and he explicitly noted that the match had to occur in all three dimensions of patch size, color diversity, and brightness.
8. Hanlon's experiments are described in Hanlon and Messenger (1996), Chiao and Hanlon (2001), Hanlon et al. (2009), and Allen et al. (2010). Cuttlefish studies are reviewed more generally in Zylinski and Osorio (2011). How cuttlefish change their color patterns is detailed in Messenger (2001) and Mäthger et al. (2008). Like all cephalopods, cuttlefish are unable to perceive color visually, but because their body pigments are similar to the colors of the environment, the pattern-matching system produces a very close resemblance in both color and pattern (Mäthger et al. 2008).
9. John Maynard Smith helped to develop the field of population genetics and was one of the first scientists to apply game theory to explain evolutionary processes. The discussion of his system for classifying animal color strategies is from Maynard Smith (1962, 1970, 1993). Color change in vertebrates and crustaceans is reviewed in Scott (1965), Bagnara and Hadley (1973), and Rao (2001). African cichlids are from Barlow (2000). Arctic species are from Caro (2005). Peppered moth developmental flexibility is from Noor,

Parnell, and Grant (2008). Additional examples are provided in Matthews and Matthews (1978), Pough (1988), Greene (1997), Langstroth and Langstroth (2000), Pianka and Vitt (2003), and Naskrecki (2005). Effects of variability and predictability are discussed in Piersma and van Gils (2011).

5. Colors in the Mind

1. As a teacher, philosopher, musician, and raconteur, Feynman was altogether unique. He combined a passion for research and a razor-sharp scientific insight with a remarkable ability to explain physics in terms that could readily be understood by nonscientists. Our account was largely inspired by Feynman (1965) and Sykes (1994). Water strider biology is in Wilcox (1995). Strictly speaking, individual molecules do not either reflect or transmit light; they just absorb a photon or they don't. What happens to photons that are not absorbed is a result of the properties of the surrounding medium. More information about visual optics is in Gregory (1978), McIlwain (1996), and Snowden, Thompson, and Troscianko (2006).
2. The role of keratin and chitin in structural coloration is discussed in Large et al. (2007) and Maia et al. (2009). Mechanisms of structural coloration are described in Simon (1971), Prum (2006), Seago et al. (2009), and Shawkey, Morehouse, and Vukusic (2009). See also Nijhout (1991), Ghiradella (1991, 1999), and Ghiradella and Butler (2009). Kinoshita (2008) has excellent color illustrations. Examples given are hummingbirds (Greenewalt, Brandt, and Friel 1960), morpho butterflies (Nijhout 1991; Kinoshita 2008), and peafowl (Yoshioka and Kinoshita 2002).
3. General introductions to biological pigments are in Vuillaume (1969), Fox (1976, 1979), Needham (1974), and Vevers (1982). The energy carried by a photon is an inverse function of the wavelength, so blue photons provide more energy than red ones. Most nonpigmented molecules are affected only by higher-energy, ultraviolet photons. Transparent animals are reviewed in Johnsen (2001). The most common glass catfish in aquarium stores is *Kryptopterus bicirrhis*. Their transparent skin and musculature have made them popular subjects for physiological investigations (e.g., Jørgensen 1991, 1992). There are several species of these odd animals in Asia and an unrelated catfish family in Africa with a similar degree of transparency.
4. Information on porphyrins is from Needham (1974), Fox (1976), and Milgrom (1997). The spectrum of chlorophyll is from Engelmann (1881)

and Blinks (1954), that of hemoglobin is from Horecker (1943) and Bowen (1949), that of hemocyanin is from Redfield (1930). Porphyrins in earthworms and snails are from Kennedy (1975); from owls and turacos are in Dyck (1992), del Hoyo, Elliott, and Sargatal (1997), and McGraw (2006).

5. Carotenoid information is from Needham (1974), Fox (1976, 1979), and Vevers (1982). Reviews of the functions of carotenoids in green plants are in Koyama (1991) and Bartley and Scolnik (1995). Carotenoid absorption spectra: xanthophylls are in Ruban (2010), and lycopene and alpha- and beta-carotene are in Miller et al. (1935). Distribution among plant species is in Goodwin (1980). Animal examples of carotenoids include canaries in Brockmann and Völker (1934), monarchs in Landrum et al. (2010), goldfish in Rothschild (1975), red-spotted newts in Davis and Grayson (2008), flamingos in Fox (1962), lady beetles in Britton et al. (1977), and common starfish in Rothschild (1975). The functions of retinol are described in Combs (1998). The universal role of retinol in vision is from Warrant and Nilsson (2006), and Lane (2009).
6. Many spider colors are due to ommochromes (Oxford and Gillespie 1998; Théry et al. 2011), which are derived in part from pigments in the eyes of their insect prey. The green in grasshoppers is insectoverdin, a combination of a yellow carotenoid with a blue bile pigment from the breakdown of porphyrins (Rowell 1971; Fuzeau-Braesch 1972). The combination of a blue structural color with yellow carotenoid to make green is common in vertebrates (Bagnara, Fernandez, and Fujii 2007; Shawkey, Morehouse, and Vukusic 2009). Among birds, only turacos, jacanas, and several species of pheasant have true green feather pigments (Dyck 1992). Guanine is a component and breakdown product of DNA. It is brilliant white and is displayed widely in the white, reflective skin cells of fishes, amphibians, and reptiles (Bagnara and Hadley 1973; Vevers 1982). The variation in chemical mechanisms among species is so great that it is never safe to assume that a given animal color is due to any particular group of pigments (Vevers 1982).
7. Reviews of melanin in biological systems are in Prota (1992) and Slominski et al. (2004). The melanin absorption spectrum is from Kollias (1995). Melanin has been identified from most major groups of animals, with the exception of spiders, which rely mainly on ommochromes (Oxford and Gillespie 1998). Fossilized dinosaur melanin granules are described in Zhang et al. (2010), Li, Gao, Meng, et al. (2012), and Li, Gao, Vinther, et al. (2012). Functions of fungal melanin are reviewed in Jacobson, Hove, and Emery (1995) and Casadevall, Rosas, and Nosanchuk (2000). Cephalopod ink has been a primary source of material for biochemical research on melanins for over a century (Prota

1992). Melanin seems to be limited to the ink sac in cephalopods; it does not contribute to their legendary flexible coloration (Messenger 2001). Melanin is the basis for darkening of insect cuticle, under both genetic and physiological control (Fuzeau-Braesch 1972; True 2003). In the vertebrate eye, melanin is concentrated in the choroid, a layer of blood vessels and connective tissue that surrounds the retina. How absorption of stray light enhances vision in brightly illuminated environments is described in Land (1997) and Land and Nilsson (2002). The role of melanin in protecting against damage from UV radiation is far from universally accepted. The various sides of the issue are discussed in Hill (1992) and Zeise, Chedekel, and Fitzpatrick (1995). Immune function and defense against infections are discussed in Burkhart and Burkhart (2005). Thermoregulation through control of melanin concentrations in the skin is well known in insects (e.g., Watt 1968; Brakefield 1987; True 2003; see also Sugumaran 2002). It is not clear whether it is important in vertebrates (Trullas, van Wyk, and Spotila 2007).

8. Melanin synthesis is outlined in Prota (1992). The relationship of melanocytes to skin and nerves is not a coincidence. All of these tissues are derived from the same cells during the development of the early embryo (Prota 1992; Yaar and Park 2012). Mechanisms of color change in lower vertebrates are described in Bagnara and Hadley (1973). Melanocytes in these organisms are often called melanophores to emphasize their dynamic capabilities. Physiological controls over melanin synthesis are reviewed in Bagnara and Hadley (1973) and Slominski et al. (2004). Genetic controls are reviewed in Protas and Patel (2008). The *Mc1r* gene regulates melanin production in a surprising number of vertebrates, including birds (Mundy 2005), mammals (Majerus and Mundy 2003), and reptiles (Rosenblum, Hoekstra, and Nachman 2004).
9. Melanin is primarily responsible for generating textures or patterns in animal coloration. See Osorio and Vorobyev (1997), Osorio, Miklósi, and Gonda (1999), Jones and Osorio (2004), and Zylinski and Osorio (2011).

6. The Beholder's Eye

1. That concealing coloration should be viewed through the eyes of appropriate species was one of Frank Beddard's many criticisms of the early coloration theorists (Beddard 1892, 89). Alfred Russel Wallace dismissed such concerns with the rather circular argument that the coloration of animals

must have evolved under the influence of their predators' visual systems, so if they appear concealingly colored to us, our vision is sufficiently similar to that of other animals (e.g., Wallace 1867, 42). Wallace was correct in the sense that there are similarities in the way human and animal visual systems perceive patterns and textures (Zylinski and Osorio 2011). The multiple origins of eyes were originally described in Walls (1967) and more recently reviewed by Salvini-Plawen and Mayr (1977). How to describe the homologies among different eyes is complicated because of the universal similarities in visual genetics and molecular biology. For a more modern, nuanced view, see Fernald (1997, 2000). Diversity of form and function is discussed in Walls (1967) and Land and Nilsson (2002). Diversity in imaging mechanisms is from Land and Fernald (1992), Land (2003), and Speiser and Johnsen (2008). *Anomalocaris* eyes are in Paterson et al. (2011). The discovery and bizarre anatomy of *Anomalocaris* is narrated by Gould (1989), with more recent interpretations by Conway Morris (1998). Presence of eyes in most Cambrian animals is from Land and Nilsson (2002). Using very conservative assumptions about possible rates of morphological change, Nilsson and Pelger (1994) calculated that a camera-type eye with a lens and a dedicated sheet of photoreceptors could evolve from flat, light-sensitive skin in 400,000 generations. At 1 generation per year, this is less than half a million years. In geological terms, it is barely a blink of an eye.

2. The retinal mechanism of vision is from Goldsmith (1990) and Dowling (2012). One would think that if the photoreceptive membrane was concentrated in one end of the cell, it should be oriented with the business end outward, toward the incoming light, sort of the way you would point a microphone toward someone who was giving a speech. But vertebrate retinas develop inside out. The membrane stack is pointed inward, toward the brain, and light has to pass through the length of the cell, as well as through a lot of other nervous tissue, to set off a signal. Goldsmith (1990) remarked that this was comparable to placing a diffusing screen between the lens and the photoreceptor array of a camera: It cannot help but degrade the image. That vertebrate eyes are not an impressive argument for intelligent design has been eloquently pointed out by Dawkins (1986) and Carroll (2006). Evolution of opsins and retinals is from Goldsmith (1990), Land and Fernald (1992), Fernald (1997, 2000), and Ödeen and Håstad (2003). The *PAX-6* gene and its evolution is described in Gehring (1998, 2005). There are even suggestions of the use of similar genetic mechanisms in both vertebrates and jellyfish, resulting in an independent evolution of a camera-type eye in both groups (Kozmik et al. 2008).

3. Low-level developmental consistency generally increases evolutionary flexibility (Carroll 2005). Ecological plasticity of eyes is discussed in Lythgoe (1979). Trade-offs between sensitivity and acuity are from Goldsmith (1990), Land and Nilsson (2002), and Dowling (2012). The multiple individual receptors in the compound eyes of arthropods vary in both acuity and sensitivity, but they do not divide readily into classes comparable to rods and cones (Land 1997; Briscoe and Chittka 2001). Dependence of color vision on multiple cone pigments is from Goldsmith (1990). Human absorption spectra are detailed in Land and Nilsson (2002), Osorio and Vorobyev (2008), and Dowling (2012). We refer to light between 430 and 440 nm as dark blue, because it includes the wavelength emitted by the blue pigment in an RGB monitor. Isaac Newton called this range of wavelengths indigo, but that term has become uncommon.
4. British marine biologist John Lythgoe was intrigued with the relationship between the structure and function of fish eyes and the nature of the ambient illumination. In his papers and his classic textbook (Lythgoe 1979), he extended the reach of his idea well beyond fishes and essentially founded the field of visual ecology, in which the functioning of the visual system is seen as an evolved response to both the spectral range and intensity of the environmental light and the demands of the species' ecology. Trichromatic vision is the minimum required for high-resolution color (Osorio and Vorobyev 2008). Honeybee absorption spectra are discussed in Osorio and Vorobyev (2008); insect trichromats in Briscoe and Chittka (2001); bird vision in Hart, Partridge, Bennett, et al. (2000), Hart, Partridge, Cuthill, et al. (2000), Håstad, Victorsson, and Ödeen (2005), Cuthill (2006), and Kram, Mantey, and Corbo (2010). Reptile tetrachromats probably include most lizards and turtles (Arnold and Neumeyer 1987; Röll 2001; Barbour et al. 2002; Dowling 2012). Fish tetrachromats are in Levine and MacNichol (1982), Neumeyer (1992), and Ward et al. (2008). Double cones are used in achromatic (that is, noncolor) tasks, such as detecting motion or differences in brightness (Cuthill 2006; Kram, Mantey, and Corbo 2010). Many tetrachromats also tweak their absorption spectra by incorporating oil droplets that filter the incoming light and shift the peak values, apparently a means of increasing sensitivity at long wavelengths (Lythgoe 1979; Goldsmith 1990). Asian swallowtails are cited in Osorio and Vorobyev (2008), and mantis shrimp are in Cronin et al. (1994). Despite the name, mantis shrimp are actually stomatopods. Typical ecology of multiple cone species is from Lythgoe (1979). Dichromatic mammal examples are from Anderson and Jacobs (1972), Jacobs and Neitz (1986), and Calderone, Reese,

and Jacobs (2003). Visual spectra of domestic cattle are from Jacobs, Deegan II, and Neitz (1998). Human color blindness is discussed in Dowling (2012). Dichromatism and trichromatism in snakes is from Sillman et al. (1997, 1999), and dichromatism in pelagic fishes is from Lythgoe (1979) and Cummings and Partridge (2001). Monochromats are discussed in Jacobs et al. (1987) and Peichl (2005).

5. Dog vision is discussed in Neitz, Geist, and Jacobs (1989) and Horowitz (2009). The situation is reversed after dusk, when the superior sensitivity of the dog's rod-dominated retina comes into its own, and human cone vision is shut down for lack of photons. Rabbits can be difficult for humans to see under dim light, even when they are close by (Thoreau 1854, 372).
6. Insects and spiders see ultraviolet very well, so to their eyes, yellow and white flowers probably look as different as bananas and blueberries do to humans (Théry et al. 2011). Natural history and timing of color change in crab spiders is from Morse (2007) and Théry (2007). The yellow color is produced by synthesizing kynurenine, one of the ommochrome pigments that provide much of the coloration in spiders (Seligy 1972). When the switch to white takes place, the yellow pigment is moved away or broken down, exposing an underlying layer of white guanine (Oxford and Gillespie 1998). Ommochromes are concentrated in the eyes of insects, and spiders must obtain them from their food. Crab spiders fed on mutant white-eyed fruit flies apparently cannot turn yellow (Fuzeau-Braesch 1972; Théry 2007). Crab spider species studied: *Misumena vatia* (Morse 2007; Théry 2007; Defrize, Théry, and Casas 2010); *Thomisus spectabilis* (Heiling et al. 2005); *Thomisus onustus* (Théry and Casas 2002; Théry et al. 2004). Spectra of spiders and flowers and color/substrate preferences are from Heiling et al. (2005). Spiders on roses are discussed in Morse (2007); their visibility to birds is from Théry and Casas (2002), Théry et al. (2004), Morse (2007), and Defrize, Théry, and Casas (2010). Lack of bird predation is in Morse (2007) and Defrize, Théry, and Casas (2010). Rapid avoidance of predators is from Morse (2007). Many female *Misumena* have bright red stripes along the sides of the abdomen. These would be invisible to the spiders' insect prey, but would be conspicuous to birds. Hinton (1976) suggested that the spiders may be aversive to birds and that the red stripes are a form of warning coloration that cannot be seen by prey species. It is a great story, but there is as yet no evidence that crab spiders taste bad. Crypticity to bees and avoidance responses are from Heiling et al. (2005, 2006) and Llandres et al. (2011), and their crypticity to flies is in Defrize, Théry, and Casas (2010). Crab spiders constantly switch from one flower to another, apparently evaluating how

rewarding the perch is and moving on if too few insects show up (Morse 2007). There is no clear indication that they use the color of the flowers in making their choice (Defrize, Théry, and Casas 2010).

7. Computational methods are detailed in Endler (1990), Vorobyev and Osorio (1998), and Endler and Mielke (2005). The problem of adaptation to multiple predator visual systems is discussed in Endler and Mappes (2004). Osorio's work on pattern perception in animals is in Osorio and Vorobyev (1997), Osorio et al. (1999), and Jones and Osorio (2004). The similarity of perceptual judgments of pattern is reviewed in Zylinski and Osorio (2011).

7. Desert Islands

1. Napoleon's invasion and the corps de savants are described in Dwyer (2008) and Strathern (2008). Tragically, Savigny lost his sight soon after coming home to France. He spent his remaining years in blindness and was never able to curate his collection of insects or write up his field notes (Lefébvre 1836; Krauss 1890). Savigny (1827) includes just his engravings.
2. Savigny's illustrations fascinated a French amateur entomologist named Alexandre Louis Lefébvre de Cérisy, who joined a medical expedition to Egypt in 1829 to see these strange creatures for himself. Lefébvre's original publication in French was in 1835, but because of its entomological importance, it was immediately translated and republished in English. Lefébvre noted: "What struck me most forcibly was the change of colour I observed in these insects according to the soil on which I found them, the tint of which they assumed in the most perfect manner; so much so, that it was only by their motions that I could distinguish them on this soil so destitute of life" (Lefébvre 1836, 68). Lefébvre was certain that these were a single species of mantids that appeared in differently colored forms, depending on the soil background, and he asked, "Do they then live in these limited spheres without wandering? [Or] can they . . . assume the colour of the soil on which they happen for a time to sojourn?" (Lefébvre 1836, 69). His question has, as yet, never been answered.
3. In his survey of the mammals and birds of the deserts of North Africa, Henri Heim de Balsac (1936) categorized each species as either matching or not matching the color of the substrate. Of 47 species of birds, he felt that 24 bore a general resemblance to the background. The mammals came out only somewhat better: Of the 50 species, 39 showed some evidence of

matching coloration. So, in one of the most barren and uniformly hued deserts in the world, roughly a third of the animals surveyed did *not* appear to be concealingly colored. Such irregularities in background resemblance caused some desert naturalists, well into the twentieth century, to question whether visual predation played any role at all in animal coloration (Meinertzhagen 1934; Mayr 1966; Hamilton 1973; Huxley 1974). Ernst Mayr was puzzled at the persistence of such skepticism. "It is somewhat disconcerting," he wrote, "that some of the naturalists who have the greatest experience with desert animals deny that selection pressure by predators could have anything to do with the formation of concealingly colored substrate races" (Mayr 1966, 329).

4. Strictly speaking, when Merriam's expedition was launched, the office he ran was called the Division of Economic Ornithology and Mammalogy. Merriam's scientific interests were not a good fit to the predominant business orientation of the Department of Agriculture, so he resorted to concealing his purposes behind vague departmental titles. It was only after an understandably sympathetic Theodore Roosevelt was elected president that Merriam's division was officially renamed the U.S. Biological Survey (Brinkley 2009). The expedition narrative is given in Merriam (1890). Merriam's classification of brown and grizzly bears is in Merriam (1918). Roosevelt strongly disagreed with Merriam's approach to bear taxonomy (Brinkley 2009). Rausch (1953, 1963) established that grizzly bears are a single, morphologically variable species. Description of Merriam's Cyclone mousetraps is in Sterling (1977).
5. The lighter form of the plains pocket mouse was discovered and described by Edward Goldman, also of the U.S. Biological Survey (Goldman 1932, 1933). Other findings are from Merriam (1890).
6. Roosevelt elaborated on his concerns in Roosevelt (1911b).
7. The account of the Chihuahuan Desert is from Jaeger (1957) and Dixon (1967). The orbital visibility of the White Sands is noted in Lindberg and Smith (1973). Pocket mice and woodrats are described in Benson (1933), Goldman (1933), Williams (1978), Braun and Mares (1989), and Monk and Jones (1996). The description of the lizards is from Smith (1943), Dixon (1967), and Rosenblum (2006); spadefoot toads are from Stroud (1949); camel crickets are from Bugbee (1942), Stroud and Strohecker (1949), and Stroud (1950).
8. Carrizozo Malpais description is from Lewis (1949) and Shields (1956). Age of the lava flow is from Dunbar (1999). Rock pocket mice are from Dice and Blossom (1937), Nachman, Hoekstra, and D'Agostino (2003), and

Hoekstra, Krenz, and Nachman (2005). Woodrats are described in Dice (1929), Hooper (1941), and Macêdo and Mares (1988); other rodents are from Benson (1932); reptiles are from Lewis (1949, 1951), Stebbins (1966), and Rosenblum (2006).

9. The Museum of Zoology at the University of Michigan was already a significant power in field biology in 1906, when Alexander Ruthven, then a graduate student in the museum, traveled through the Southwest collecting reptiles. Among his trophies were the first examples of the three white lizard species from the Tularosa gypsum dunes. The curator of mammals at the Michigan museum, Lee Dice, led a subsequent expedition to collect rodents in the same area in 1927 (Dice 1929). His description of black and white mice from the Tularosa, combined with Ruthven's (1907) earlier discoveries, led to thorough surveys of sand and lava areas throughout the Southwest. Annie Alexander's influence at Berkeley was immense: Her donations originally established and funded the Berkeley museum, she recommended the first director, Joseph Grinnell, and she was an avid collector and field naturalist (Stein 2001). Seth Benson, then a graduate student at the Berkeley museum, undertook a comprehensive study of the area (Benson 1932, 1933). Benson's work was followed by more mammal studies by Michigan researchers (Bradt 1932; Dice and Blossom 1937; Hooper 1941; Blair 1943). Mammal surveys are from Benson (1933), Hooper (1941), and Blair (1943). The reptile surveys are from Lewis (1949, 1951). Benson emphasized, "[C]oloration paralleling that of the White Sands or of the malpais appears only among those mammals which tend to be isolated on those areas" (Benson 1933, 46). He concluded that "[n]atural selection for concealing coloration, and isolation, appear to have been the chief factors in the development of the white race of pocket mouse on the White Sands and the dark races of mammals on the black lava bed" (Benson 1933, 2).

8. Flowing Genes

1. Wallace's years in the Malay Archipelago are described in Daws and Fujita (1999). He described his travels as "wanderings among the largest and the most luxuriant islands which adorn our earth's surface" (Wallace 1869, 455). From Sarawak on the island of Borneo, Wallace wrote his early statement on evolution, "On the Law Which Has Regulated the Introduction on New Species" (Wallace 1855). Later, from the island of Ternate in

Indonesia, he wrote the short essay that, with Darwin's work, laid out the principle of evolution by natural selection (Darwin and Wallace 1858). On the Galápagos, Darwin wrote, "[S]everal of the islands possess their own species of the tortoise, mocking-thrush, finches, and numerous plants, these species having the same general habits, occupying analogous situations, and obviously filling the same place in the natural economy of this archipelago" (Darwin 1845, 397). The oldest islands in the Galápagos archipelago are about five to ten million years of age. The youngest islands, Isabela and Fernandina, are still volcanically active (Jackson 1993).

2. Mayr's 'island hypothesis' is presented in Mayr (1942, 1967) and Mayr and Diamond (2001). Mayr and Diamond (2001) investigated monarch flycatchers on the Solomon Islands in the South Pacific. Although some distinctive forms were on islands close to one another, there were also entire strings of islands all inhabited by the same form of monarch flycatcher. The linear distance between islands is not the only factor involved in the divergent evolution of these birds. Although migration is restricted by oceanic barriers, several flycatcher subspecies have evolved additional, behavioral barriers to gene flow. Biases in mate selection substantially reduce the chance that a migrant from another island will be able to breed with a local bird (Filardi and Smith 2008; Uy, Moyle, and Filardi 2009), and such biases can evolve exceedingly rapidly (Uy et al. 2009). Island evolution has recently been reviewed by Whittaker and Fernández-Palacios (2007) and Losos and Ricklefs (2009). The limited dispersal of rock pocket mice is from Findley et al. (1975) and Hoffmeister (1986); fringe-toed lizards are from Norris (1958).
3. The details of this evolutionary mechanism were developed by Sewall Wright (1978). Wright's primary focus was on the evolution of small, isolated populations in fragmented habitats. An introduction to population genetic concepts is in Conner and Hartl (2004); the full-blown treatment is Hartl and Clark (2007). Mayr and Wright converged on a similar account of evolution, but Mayr's view of population theory was uniformly scornful, asserting that Wright and his colleagues were practicing "beanbag genetics." For various perspectives on their relationship, see Provine (1986, 2004). Ridley (2004) provides a clear explanation of the modern synthetic view of genetics and natural selection and the relationships among genetic drift, natural selection, and gene flow. Consequences of the three population processes are not always clean and simple. When populations are very small and very isolated, the random dance of genetic drift can result in the extinction, purely by chance, of some of the alternative forms. Genetic uniformity in a very small population can therefore be produced by chance, even in the

face of natural selection (Wright 1978). The same caveat can be applied to gene flow. Gene flow slows evolutionary innovation, but it can also enhance the odds of avoiding local extinction. One of the most elegant examples of natural selection in the wild is the work on medium ground finches and cactus finches on the Galápagos Islands by Rosemary and Peter Grant. Accounts are in Grant and Grant (1989) and Grant (1986, 1998), as well as in *Beak of the Finch,* a Pulitzer Prize–winning account by Weiner (1994). In these studies, gene flow produces populations with a mixture of traits. Such hybrids are generally healthier than animals from inbred, isolated populations, and they are better able to survive abrupt changes in their environment or food supply.

4. Pocket mice are members of the Heteromyidae, along with kangaroo rats and some less familiar creatures (Walker 1975; Genoways and Brown 1993). Pocket mouse biology is reviewed by Eisenberg (1963), Prakash and Ghosh (1975), and Randall (1993). The role of these rodents in desert communities is discussed in Kotler and Brown (1988) and Whitford (2002). Pocket mouse diets are from Reichman (1975). Cheek pouch anatomy is described in Nikolai and Bramble (1983) and Ryan (1986). Vulnerability to predators is documented in Schmidly (1977) and Wilson and Ruff (1999).
5. The habitat preferences of rock pocket mice are described in Findley et al. (1975) and Hoffmeister (1986). Dice and Blossom's studies are in Dice and Blossom (1937), and experiments with owls are described in Dice (1947) and Kaufman (1974).
6. How the researchers worked out the coat-color gene in rock pocket mice is described in Nachman, Hoekstra, and D'Agostino (2003). Control of eumelanin synthesis by *Mc1r* is in Barsh (1996). Mutations in *Mc1r* can be constitutive, meaning that they fail to limit eumelanin production and produce a very dark coloring, or they can be nonfunctional, resulting in a light coat color. Many species of mammals show multiple color phases due to variation in *Mc1r,* including cats (Eizirik et al. 2003), bears (Ritland, Newton, and Marshall 2001), squirrels (McRobie, Thomas, and Kelly 2009), and even mammoths (Römpler et al. 2006). Hoekstra and Nachman tested dark mice from other lava beds in southern New Mexico and found that their color was not due to an *Mc1r* mutation. The New Mexican mice apparently evolved their coloration independently via changes in some other gene (Hoekstra and Nachman 2003).
7. Most of the work on population genetics is from Hoekstra, Drumm, and Nachman (2004). The researchers compared sequences from two mitochondrial DNA genes to compute the phylogeny and then used a statisti-

cal modeling program to obtain their population estimates (Hoekstra and Nachman 2005). Description of the lava field is from Green and Short (1971) and Gutmann and Sheridan (1978). The final results are surprisingly similar to those from the peppered moth studies, where the light form also seemed more vulnerable to predation in the nonmatching environment, and the intensity of selection on Hoekstra's mice is fully comparable to what J. B. S. Haldane estimated for peppered moths during the worst years of the coal smoke pollution (Nachman 2005).

8. Our perspective on lizard biology is from Pianka (1986) and Ayal (2007). The history and predator ecology of the Tularosa is from Degenhardt, Painter, and Price (1996), Des Roches et al. (2011), and Robertson et al. (2011). Predators of the Tularosa basin lizards include northern mockingbirds, common ravens, Swainson's hawks, tyrant flycatchers, greater roadrunners, longnose leopard lizards, collared lizards, western rattlesnakes, coachwhips, and gopher snakes.
9. The primary study is in Rosenblum (2006). See also Rosenblum, Hoekstra, and Nachman (2004) and Rosenblum et al. (2010). Lesser earless males have a pair of vertical black bars, edged in blue, along their sides; male fence lizards are blotched with dark blue along the sides of the belly and throat; and whiptail males have bright blue heads (Degenhardt, Painter, and Price 1996; Robertson and Rosenblum 2009). Reproductive consequences of the genetic differences are explored in Rosenblum (2008), Robertson and Rosenblum (2009), and Rosenblum and Harmon (2010).

9. Telltale Signs

1. The role of learning in predation is reviewed in Krebs (1973) and Curio (1976). The universality of concealment in the tropics and its effects on predator cognition are discussed in Robinson (1979, 1990). The ecology of the callithricid monkeys (tamarins and marmosets) in the New World tropics is reviewed by Sussman and Kinzey (1984). Hunting behavior in Panamanian red-crested tamarins is described in Robinson (1970) and Caine (1996), though it is very similar to that of other members of the family (e.g., Passos and Keuroghlian 1999; Raboy and Dietz 2004; Schiel et al. 2010).
2. Nikolaas Tinbergen pioneered the study of behavioral evolution, and he directly influenced an entire generation of color researchers, many of whom are highlighted in this book. He advised Michael Robinson on his tamarin

studies, Leen de Ruiter on his caterpillar camouflage work, and Harvey Croze on his studies of visual search in carrion crows. He was advisor to Marian Stamp Dawkins in her experiments on searching image, as well as to the evolutionary biologist, Richard Dawkins. He was older brother to Luuk Tinbergen, who first explored the evolution of prey polymorphism, and he assisted Bernard Kettlewell in his behavioral experiments on moth predation. In 1973, Niko Tinbergen shared the Nobel Prize in Physiology or Medicine with Karl von Frisch and Konrad Lorenz for their work on the study of social behavior and instinct. One of Tinbergen's most popular books is *Curious Naturalists,* which describes his work with Kettlewell and de Ruiter in considerable detail (Tinbergen 1958). Tinbergen's impact on the field is reviewed in Dawkins, Halliday, and Dawkins (1991).

3. Robinson's studies on concealing coloration are reviewed in Robinson (1969, 1991). The experiments on red-crested tamarins, conducted at the Smithsonian Tropical Research Institute in Panama, are described in Robinson (1970), and their implications are discussed in Robinson (1969). Robinson eventually became director of the National Zoo at the Smithsonian Institution in Washington, D.C. Many tropical phasmids even have a special notch in the femurs of their forelegs, which allows the legs to lock together over the top of the head and completely conceal it from view (Robinson 1969). The two species that Robinson used were *Metriophasma diocles* and *Pseudophasma menius.* Drawing their legs in and dropping to the ground is not the only escape option available to phasmids. Many species are good fliers and can take wing if they are harassed. A comprehensive account of the Phasmida is in Otte and Brock (2003).
4. Discrimination learning has been a core focus of the field of comparative psychology, dating back to the work of Spence (1936), who originally theorized that responses to positively rewarding stimuli would be learned more rapidly when they were directly contrasted to unrewarding ones (Riley 1968; Sutherland and Mackintosh 1971). The expertise of marmosets at discrimination learning has been extensively documented (Miles and Meyer 1956; Cotterman, Meyer, and Wickens 1956; Stellar 1960; Roberts, Robbins, and Everitt 1988). Kea foraging is described in Diamond and Bond (1999). The acquisition of foraging in marmosets is greatly influenced by the behavior of older individuals (Schiel and Huber 2006; Souto et al. 2007; Rapaport and Brown 2008; Schiel et al., 2010). Aviary studies of foraging marmosets were conducted by Menzel and Menzel (1979) and Menzel and Juno (1982).
5. Goodwin (1986) provides the best species description of the natural history and behavior of Eurasian jays. Also see Madge and Burn (1994). De Ruiter's

experiments (de Ruiter 1952) are described in engaging detail in Tinbergen (1958). The caterpillars used were *Ennomos alniaria,* which feeds on birch trees; *Biston hirtaria,* which feeds on plum trees; and *Biston strataria,* which feeds on willows. Jay taxonomy is in Ericson et al. (2005); blue jay natural history is in Goodwin (1986) and Tarvin and Woolfenden (1999). Discrimination learning in blue jays has been studied by Kamil et al. (1977), Pietrewicz and Kamil (1977), and Bond and Kamil (1999).

6. The four criteria are described in Tinbergen (1965) and restated with examples in Robinson (1969). Robinson (1981) emphasized the crucial function of abundant, unrewarding model objects. His title, "A stick is a stick and not worth eating: on the definition of mimicry," is a capsule summary of the mechanism. See Naskrecki (2005) for a discussion and extensive photographs. Tinbergen, Impekoven, and Franck (1967) conducted an experiment on how changing the spacing between concealingly colored eggs affected the predation success of carrion crows. They found results that were consistent with Tinbergen's prediction of the effects of low density on camouflaged prey.

10. Psychology of Search

1. Fleming rightly emphasized the complexity of the coral reef background, rather than the cryptic coloration of the fish James Bond was seeking, as the main justification for focused visual search (Fleming 1960). The reconstruction of our visual world is elegantly described by Hoffman (1998). Attention has been a primary concern of cognitive psychologists for well over a century. William James, the father of American psychology, notoriously remarked that "[e]very one knows what attention is" (James 1890, 403), but rather than a single cognitive process, attention has proved to be an exceedingly diverse and complex set of vaguely related phenomena (Parasuraman and Davies 1984; Pashler 1998). An accessible account of visual attention is in Humphreys and Bruce (1989). More technical reviews are in Wolfe (1998), Chun and Wolfe (2001), and Carrasco (2011). Stimulus "pop-out" is most apparent when search is focused on a single, conspicuous feature, like a spot of international orange (Treisman and Gelade 1980). The down side of attention is called inattentional blindness, a vivid expression of how costly visual attention can be (Simons 2000; Mack and Rock 2000).
2. The most detailed conceptual model of guided search is in Wolfe (2007). See also Itti and Koch (2000, 2001). Psychologists refer to this form of bottom-up

guidance as repetition priming because repeating a search for the same item over a series of experimental trials appears to prime the attention mechanism the way one might prime a water pump (Kristjánsson and Campana 2010). The searching image anecdote is from von Uexküll (1934). The role of searching images in natural visual search is reviewed in Langley (1996) and Bond (2007). Searching images really do enhance visual sorting: More difficult sorting tasks, where the items closely resemble one another, are conducted in longer sequential runs with fewer switches between item types (Bond 1982).

3. Luuk was the younger brother of Nikolaas Tinbergen. Luuk Tinbergen's field study of great tits and caterpillars is Tinbergen (1960). Factors that produce switching between prey types are described in Holling (1965) and Murdoch and Oaten (1974). A more recent review is Sinervo and Calsbeek (2006). Studies of animal visual search in the field, such as Croze's (1970) experiments on carrion crows and Murton's (1971) investigations of wood pigeons, make clear that foraging behavior in the wild is a rich and varied phenomenon, but they also illustrate the challenges it poses for testing causal hypotheses. In Bond's (1983) study, pigeons were given samples of varying proportions of familiar, natural food grains spread at random over a complex gravel background. When the grains strongly resembled the gravel, the birds all switched back and forth between grain types. The effect disappeared when the grains were conspicuous, which was clear evidence of the use of searching images. Similar results have been obtained in other free choice experiments (Cooper 1984; Tucker 1991; Reid and Shettleworth 1992; Cooper and Allen 1994; Langley et al. 1995).
4. Marian Stamp Dawkins was another of Niko Tinbergen's students. Her experiments are described in Dawkins (1971a, 1971b). The theoretical basis for her "runs effect" was given by McNair (1980). Pigeon searching image studies include Bond (1983), Blough (1989, 1991, 1992), Bond and Riley (1991), Langley et al. (1995), and Langley (1996). Zentall and Riley have reviewed the evidence for visual attention in animals, focusing on the underlying cognitive process (Zentall and Riley 2000; Zentall 2005). Kamil's original blue jay and underwing moth experiments were Pietrewicz and Kamil (1977, 1979). Searching image studies using digital moths include Bond and Kamil (1998, 1999, 2006) and Dukas and Kamil (2000, 2001). They were reviewed in Bond (2007). Animals can locate multiple conspicuous items using nonattentive search mechanisms (Dukas 2002; Kamil and Bond 2006; Bond 2007), but fully camouflaged targets cannot be successfully detected with a nonattentive search (Wolfe 1994; Wolfe et al. 2002).

5. It seems likely that in a perceptually demanding search, only one stimulus representation can be held in memory at a time (Langley 1996; Gorea and Sagi 2000; Kristjánsson and Campana 2010), so it is not possible to conduct a simultaneous focal search for more than one distinctive target. The porosity of visual attention has been remarked on in animal studies (Blough 1992; Bond and Kamil 1999), and it is the basis for an entire subfield of the human visual search literature (reviewed in Rauchenberger 2003 and Boot, Kramer, and Becic 2006). The gradual attenuation of searching images over time has been noted by Bond (1983), Langley et al. (1995), and Langley (1996). The idea that persistence in attending is based on time expectancy ties into research on how rapidly a predator should move through the environment (Gendron 1986; Gendron and Staddon 1983, 1984) or how long it should survey discrete locations (Bond and Riley 1991; Endler 1991; Getty and Pulliam 1991). Getty has helped to clarify the complementary roles of perceptual and tactical constraints in the allocation of search effort (Getty, Kamil, and Real 1987; Getty and Pulliam 1991, 1993).
6. The psychological literature calls the disruption of focal search attentional capture. How reliably a particular stimulus can capture attention is a complicated issue that goes beyond just the strength of the original attentional focus (Turatto and Galfano 2000; Horstmann 2002; Morrone, Denti, and Spinelli 2002; Rauchenberger 2003; Boot, Kramer, and Becic 2006). Some studies have found continuing increases in accuracy with progressively longer runs of the same stimulus type, suggesting that even the initial allocation of attention may be a graded phenomenon (Maljkovic and Martini 2005; reviewed in Kristjánsson and Campana 2010). All these results lead to the idea of a smooth fluctuation among alternative perceptual states (Falmagne 1965; Schweickert 1993; Maljkovic and Martini 2005). This model would help to explain the occurrence of strong searching image effects under conditions of continuously variable prey appearance (Bond and Kamil 2002, 2006; Bond 2007). Unlike bright colors or patterns, disruption due to animate motion is able to override nearly any focal visual search (Abrams and Christ 2003; Franconeri and Simons 2003; Pratt et al. 2010; Sunny and von Mühlenen 2011).
7. A key to identifying katydids, either by sound or by appearance, is in Alexander, Pace, and Otte (1972). Himmelman's (2011) lovely account of his life with katydids and crickets includes his first visual encounter with bush katydids.

11. Distinctively Different

1. In the introduction to his field guide, Sibley reminds readers, "You should not expect any bird seen in the field to match exactly the picture of its species in the field guide" (Sibley 2003, 11). Field identification of sandhill and whooping cranes is from Sibley (2003).
2. Eclectus parrots are described in Higgins (1999), Forshaw (2002), and Heinsohn, Legge, and Endler (2005). Wallace (1865, 1869) describes examples of swallowtails in Indonesia, the most spectacular of which is the great mormon. The multiple morphs of swallowtails are often Batesian mimics of other butterflies (Wickler 1968; Kunte 2009)
3. E. B. Ford (1975) originated the term *polymorphism*. Color polymorphism has been extensively studied in banded snails (Cain and Sheppard 1954; Clarke et al. 1978; Cook 1986, 1998; Ożgo 2011), but it is also found in other mollusks (Atkinson and Warwick 1983; Whiteley, Owen, and Smith 1997). There are many records of polymorphic camouflaged spiders (Oxford and Gillespie 1998; Oxford 2009), crustaceans (Palma and Steneck 2001; Devin et al. 2004; Hargeby, Johansson, and Ahnesjö 2004), and vertebrates (Benson 1933; Norris and Lowe 1964; Endler 1978; Hoffman and Blouin 2000). Most insect orders include polymorphic species (Richards 1961). Polymorphism occurs throughout the short-horned grasshoppers (Dearn 1990) and grouse locusts (Nabours, Larson, and Hartwig 1933), where many genera are so variable that color is wholly inadequate as a basis for classification (Rowell 1971). Many mantids (Edmunds 1972, 1976) and katydids (Naskrecki 2005) are polymorphic, as are water boatmen (Popham 1941), walking sticks (Sandoval 1994a), sand-burrowing beetles (Harris and Weatherall 1991), and spittlebugs (Brakefield 1990b; Halkka and Halkka 1990). Perhaps the best-researched instances of color polymorphism are found among Geometrid and Noctuid moths, which are generally cryptically colored and frequently polymorphic (Kettlewell 1973; Sargent 1976). Underwing moths are detailed in Barnes and McDunnough (1918).
4. Environmentally based polymorphisms were reviewed by Hedrick (1986, 2006). Grasshopper habitat choices are in Eterovick, Figueira, and Vasconcellos-Neto (1997) and Pellissier et al (2011). The wrangler grasshopper was studied by Gillis (1982); the morphs of the white underwing moth are described in Barnes and McDunnough (1918), and its habitat choices were determined by Sargent (1981). Sandoval (Sandoval and Crespi 2008) traced the phylogenetic distribution of striping in phasmids. Her work in

the California chaparral on the phasmid, *Timema cristinae*, is detailed in Sandoval (1994a, 1994b). The habitat choice of these insects was researched by Nosil, Sandoval, and Crespi (2006) and Nosil (2007). Green and brown polymorphisms are described in Poulton (1890), Edmunds (1976), Owen (1980), Dearn (1990), and Wente and Phillips (2003). Dark and light forms are even more common (Kettlewell 1973; Majerus 1998).

5. Edward Poulton, in his classic work *The Colours of Animals* (1890) anticipated the effects of searching images. He described woodland caterpillars that occurred in both brown and green forms and suggested that the coloration probably served to reduce the efficiency of predatory search. Predators hunting for polymorphic caterpillars, he wrote, "have a wider range of objects for which they may mistake the larvae, and the search must occupy more time, for equivalent results, than in the case of other species which are not dimorphic" (Poulton 1890, 47). Clarke's innovative interpretation of searching images was called apostatic selection, a higher relative mortality among more abundant prey types that serves to stabilize prey polymorphism (Clarke 1962a, 1962b, 1969). Frequency-based polymorphism is reviewed in Bond (2007) and Punzalan, Rodd, and Hughes (2005). Examples of polymorphisms consistent with hunting by searching image are obscure grouse locusts (*Tetrix arenosa*; Nabours, Larson, and Hartwig 1933), banded snails (Cain and Sheppard 1954), and little bride underwing moths (*Catocala micronympha*; Barnes and McDunnough 1918). Brittle stars were studied by Moment (1962). The beach clam examples include *Donacilla cornea* in Europe (Owen and Whiteley 1988; Whiteley, Owen, and Smith 1997) and *Donax variabilis* in North America (Moment 1962). The fieldwork on carrion crows was conducted by Croze (1970). There are numerous other examples of massive polymorphism, including California meadow spiders (Oxford 2009), guppies (Endler 1978), and the army cutworms that emerge as adults in huge numbers in the late summer (Common 1954; Owen and Whiteley 1986, 1989).
6. Trinidad lies just a few kilometers off the coast of Venezuela. The Northern Range mountains are actually the easternmost extension of the Andes. The ecology and evolution of guppies (*Poecilia reticulata*) have spawned an enormous scientific literature. A recent review of this material is in Magurran (2005). Courtship in guppies was originally described by Baerends, Brouwer, and Waterbolk (1955); effects of color on mate choice are extensively reviewed in Houde (1997) and Brooks (2002). Predation on guppies is discussed in Endler (1978) and Magurran (2005). The characin is the twospot astyanax (*Astyanax bimaculatus*), the killifish is the leaping

guabine (*Rivulus hartii*), the cichlid is the millet (*Crenicichla alta*), and the freshwater shrimp is *Macrobrachium crenulatum*. The initial fieldwork was described in Endler (1978), with subsequent experimental studies in Endler (1980, 1983). There have been other remarkable experiments done on the guppy system. For example, Endler found that guppies in greenhouse pools change their courtship behavior, as well as their coloration, in response to predation pressure (Endler 1987). And in the field, guppy color patterns turn out to be more conspicuous to fish vision at the times and places when courtship is most frequent and less conspicuous at times and places of maximum predation (Endler 1991).

7. The frequency dependence study is described in Olendorf et al. (2006). There are additional factors that could help to maintain massive polymorphism in male guppies. One entertaining possibility is that female guppies could be accentuating the effects of predators by tending to choose rare and/or unfamiliar males as mates (Hughes et al. 1999; Kelley, Graves, and Magurran 1999). The criteria that female guppies use to choose mates seem remarkably chaotic (Brooks and Endler 2001; Brooks 2002), but the attraction to unusual males is statistically very compelling, and it would certainly serve to promote polymorphism.
8. The polymorphism experiments with jays and digital moths are described in Bond and Kamil (1998, 2002, 2006) and reviewed in Bond (2007).

12. Limits to Invisibility

1. Reviews of multicomponent defensive strategies are in Edmunds (1974), Evans and Schmidt (1990), Caro (2005), and Eisner, Eisner, and Siegler (2005). The *Basiceros* ants are described in Wilson and Hölldobler (1986) and Hölldobler and Wilson (1990).
2. Aristotle was intrigued by the idea of fluidly changeable coloration. In his *Historia Animalium* in 350 BCE, he discussed the ability of chameleons to alter their appearance (Aristotle 1910, § 503a), and he described in detail the behavior of the octopus, which "seeks its prey by so changing its colour as to render it like the colour of the stones adjacent to it" (Aristotle 1910, § 622a).
3. The dark colors of desert birds were mentioned in Al-Jāḥiẓ (1969); those of cold-water aquatic mammals were discussed by Albertus Magnus (Albertus Magnus 1999). The scientific revolution is briefly outlined in Henry (2008), its effects on Enlightenment thought are discussed in Israel (2001), and its

focus on universal principles in biology is elaborated on in Greene (1959). Erasmus Darwin used the terms *final* and *efficient causation* as Aristotle defined them. A final cause is essentially the function of the colors; the efficient cause is the biological mechanism.

4. Wallace's foundational paper was Wallace (1867). Wallace proposed additional arguments against environmental causation in Wallace (1891). Beddard's account of coloration is in Beddard (1892).
5. Poulton is something of a bridging figure in evolutionary studies. He was a close friend of Wallace in his early career and a colleague and collaborator with J. B. S. Haldane, R. A. Fisher, and E. B. Ford in his old age (Davis 2009). Poulton's biography is in Evans (1965). Poulton summarized his work on protective coloration in Poulton (1890). Color strategies that were conditioned on environmental cues were of widespread interest in the 1890s. They had been investigated by Poulton's colleague, Raphael Meldola, an entomologist and expert in color chemistry (Travis 2010), and they were crucial to the ideas of August Weismann, one of the most influential evolutionary theorists of the nineteenth century (Weismann 1882).
6. Beddard understood the difference between evolution of fixed and variable features and questioned the conditions that would be necessary to select for variability. He wrote that in the case of flexible coloration "the action of natural selection here must be quite different from the action of natural selection in producing fixed resemblances to the environment" (Beddard 1892, 139). His explanation foreshadowed modern concerns with the evolution of phenotypic plasticity: "[T]he variable pupa may be supposed to have been, as many pupae still are, originally uninfluenced by light; the acquirement of such a susceptibility, being of manifest use, was favoured by natural selection. In fact, we assume in this case not a positive change of any kind, but the acquirement of the capability of change when necessary" (Beddard 1892, 140).
7. Because of their agricultural impact, desert locusts (*Schizocerca gregaria*) have been extensively studied for well over a century. Behavioral, hormonal, and neurobiological components of locust transformation are addressed in Applebaum and Heifetz (1999), Hassanali, Njagi, and Bashir (2005), and, most extensively, Anstey et al. (2009). A detailed review of this material is in Simpson and Sword (2009), and the locust work is placed in a more general context of insect conditional strategies in Simpson, Sword, and Lo (2011). Current ideas on plasticity in insects are reviewed in Whitman and Ananthakrishnan (2009); see also Rowell (1971), Edmunds (1976), Greene (1996), and Leimar (2009). Phenotypic plasticity provided the original core

of what has now expanded into the science of evolutionary development (evo-devo), one of the most vigorous and exciting of the new directions of evolutionary theory. See Carroll (2005) for a general introduction, Brakefield (2006) for a more technical review, and Fusco (2012) for a detailed look at the most current research.

8. The lingering influence of extinct species on modern natural history is discussed in Diamond (1990) and Barlow (2002). Kakapo (*Strigops habroptilus*) natural history is reviewed in Higgins (1999). See also Elliott, Merton, and Jansen (2001), Eason et al. (2006), and Diamond et al. (2006). Haast's eagle (*Harpagornis moorei*) is described in Worthy and Holdaway (2002) and Bunce et al. (2005).

References

Abrams, R.A., and S.E. Christ. 2003. Motion onset captures attention. *Psychological Science* 14: 427–32.

Adams, Judith A. 1960. A contribution to the biology and postlarval development of the sargassum fish, *Histrio histrio* (Linnaeus), with a discussion of the *Sargassum* complex. *Bulletin of Marine Science of the Gulf and Caribbean* 10: 55–82.

Agassiz, Alexander. 1888. *A Contribution to American Thalassography: Three Cruises of the United States Coast and Geodetic Survey Steamer "Blake," in the Gulf of Mexico, in the Caribbean Sea, and Along the Atlantic Coast of the United States, from 1877 to 1880.* New York: Houghton Mifflin.

Albertus Magnus. 1999. *On Animals: A Medieval Summa Zoologica.* Translated by K.F. Kitchell Jr. and I.M. Resnick. Baltimore, MD: Johns Hopkins University Press.

Alexander, R.D., A.E. Pace, and D. Otte. 1972. The singing insects of Michigan. *Great Lakes Entomologist* 5: 33–69.

Allen, David E. 1976. *The Naturalist in Britain: A Social History.* London: Allen Lane.

Allen, J.J., L.M. Mäthger, A. Barbosa, K.C. Buresch, E. Sogin, J. Schwartz, C. Chubb, and R.T. Hanlon. 2010. Cuttlefish dynamic camouflage: Responses to substrate choice and integration of multiple visual cues. *Proceedings of the Royal Society* B 277: 1031–39.

Anderson, D.H., and G.H. Jacobs. 1972. Color vision and visual sensitivity in the California ground squirrel (*Citellus beecheyi*). *Vision Research* 12: 1995–2004.

Anstey, M.L., S.M. Rogers, S.R. Ott, M. Burrows, and S.J. Simpson. 2009. Serotonin mediates behavioral gregarization underlying swarm formation in desert locusts. *Science* 323: 627–30.

Applebaum, S.W., and Y. Heifetz. 1998. Density-dependent physiological phase in insects. *Annual Review of Entomology* 44: 317–41.

Aristotle. 1910. *The Works of Aristotle IV: Historia Animalium.* Translated by D'Arcy W. Thompson. Oxford, UK: Oxford University Press.

Arnold, K., and C. Neumeyer. 1987. Wavelength discrimination in the turtle *Pseudemys scripta elegans. Vision Research* 27: 1501–11.

Arnold, R.J., and T.W. Pietsch. 2012. Evolutionary history of frogfishes (Teleostei: Lophiiformes: Antennariidae): a molecular approach. *Molecular Phylogenetics and Evolution* 62:117–29.

Atkinson, W. D., and T. Warwick. 1983. The role of selection in the colour polymorphism of *Littorina rudis* Maton and *Littorina arcana* Hannaford-Ellis (Prosobranchia: Littorinidae). *Biological Journal of the Linnean Society* 20: 137–51.

Ayal, Yoram. 2007. Trophic structure and the role of predation in shaping hot desert communities. *Journal of Arid Environments* 68: 291–98.

Baerends, G. P., R. Brouwer, and H. T. Waterbolk. 1955. Ethological studies on *Lebistes reticulatus* (Peters). I. An analysis of the male courtship pattern. *Behaviour* 8: 249–334.

Bagnara, J. T., P. J. Fernandez, and R. Fujii. 2007. On the blue coloration of vertebrates. *Pigment Cell Research* 20: 14–26.

Bagnara, J. T., and M. E. Hadley. 1973. *Chromatophores and Color Change: The Comparative Physiology of Animal Pigmentation.* Englewood Cliffs, NJ: Prentice-Hall.

Baker, Richard A. 1996. The Great Gun of Durham—Canon Henry Baker Tristram, F.R.S. (1822–1906). An outline of his life, collections and contribution to natural history. *Archives of Natural History* 23: 327–41.

Barber, Lynn 1980. *The Heyday of Natural History: 1820–1870.* Garden City, NY: Doubleday.

Barbour, H. R., M. A. Archer, N. S. Hart, N. Thomas, S. A. Dunlop, L. D. Beazley, and J. Shand. 2002. Retinal characteristics of the ornate dragon lizard, *Ctenophorus ornatus. Journal of Comparative Neurology* 450: 334–44.

Barlow, Connie C. 2002. *The Ghosts of Evolution: Nonsensical Fruit, Missing Partners, and other Ecological Anachronisms.* New York: Basic Books.

Barlow, George W. 2000. *The Cichlid Fishes: Nature's Grand Experiment in Evolution.* Cambridge, MA: Perseus Publishing.

Barnes, W., and J. McDunnough. 1918. Illustrations of the North American species of the genus *Catocala. Memoirs of the American Museum of Natural History, New Series* 3, Part 1: 1–47.

Barsh, Gregory S. 1996. The genetics of pigmentation: From fancy genes to complex traits. *Trends in Genetics* 12: 299–305.

Bartley, G. E., and P. A. Scolnik. 1995. Plant carotenoids: Pigments for photoprotection, visual attraction, and human health. *The Plant Cell* 7: 1027–38.

Bates, Henry W. 1862. Contributions to an insect fauna of the Amazon Valley: Lepidoptera: Heliconidae. *Transactions of the Linnean Society of London* 23: 495–566.

———. 1864. *A Naturalist on the River Amazons: A Record of Adventures, Habits of Animals, Sketches of Brazilian and Indian Life, and Aspects of Nature under the Equator, during Eleven Years of Travel.* London: John Murray.

Beddard, Frank E. 1892. *Animal Coloration: An Account of the Principal Facts and Theories Relating to the Colours and Markings of Animals*. London: Swan Sonnenschein & Co.

Behrens, Roy R. 2002. *False Colors: Art, Design and Modern Camouflage*. Dysart, IA: Bobolink Books.

Benson, Seth B. 1932. Three new rodents from lava beds of southern New Mexico. *University of California Publications in Zoology* 38: 335–44.

———. 1933. Concealing coloration among some desert rodents of the southwestern United States. *University of California Publications in Zoology* 40: 1–70.

Berry, Robert J. 1990. Industrial melanism and peppered moths (*Biston betularia* [L.]). *Biological Journal of the Linnean Society* 39: 301–22.

Blair, W. Frank. 1943. Ecological distribution of mammals in the Tularosa Basin. *Contributions from the Laboratory of Vertebrate Biology, University of Michigan* 20: 1–24.

Blechman, Hardy. 2004. *Disruptive Pattern Material: An Encyclopedia of Camouflage*. Buffalo, NY: Firefly Books.

Blinks, Lawrence R. 1954. The photosynthetic function of pigments other than chlorophyll. *Annual Review of Plant Physiology* 5: 93–114.

Blough, Patricia M. 1989. Attentional priming and visual search in pigeons. *Journal of Experimental Psychology Animal Behavior Processes* 15: 358–65.

———. 1991. Selective attention and search images in pigeons. *Journal of Experimental Psychology Animal Behavior Processes* 17:292–98.

———. 1992. Detectability and choice during visual search: Joint effects of sequential priming and discriminability. *Animal Learning and Behavior* 20: 293–300.

Bond, Alan B. 1982. The bead game: Response strategies in free assortment. *Human Factors* 24: 101–10.

———. 1983. Visual search and selection of natural stimuli in the pigeon: The attention threshold hypothesis. *Journal of Experimental Psychology: Animal Behavior Processes* 9: 292–306.

———. 2007. The evolution of color polymorphism: Crypticity, searching images, and apostatic selection. *Annual Review of Ecology, Evolution and Systematics* 38: 489–514.

Bond, A. B., and A. C. Kamil. 1998. Apostatic selection by blue jays produces balanced polymorphism in virtual prey. *Nature* 395: 594–96.

———. 1999. Searching image in blue jays: Facilitation and interference in sequential priming. *Animal Learning and Behavior* 27: 461–71.

———. 2002. Visual predators select for crypticity and polymorphism in virtual prey. *Nature* 415: 609–14.

———. 2006. Spatial heterogeneity, predator cognition, and the evolution of color polymorphism in virtual prey. *Proceedings of the National Academy of Sciences* 103: 3214–19.

Bond, A. B., and D. A. Riley. 1991. Searching image in the pigeon: A test of three hypothetical mechanisms. *Ethology* 87: 203–24.

Boot, W. R., A. F. Kramer, and E. Becic. 2006. Capturing attention in the laboratory and the real world. In *Attention: From Theory to Practice,* edited by A. F. Kramer, D. A. Wiegmann, and A. Kirlik, 27–44. New York: Oxford University Press.

Bortolotti, Gary. 2006. Natural selection and coloration: Protection, concealment, advertisement, or deception? In *Function and Evolution,* edited by G. E. Hill and K. C. McGraw, 3–35. Vol. 2 of *Bird Coloration.* Cambridge, MA: Harvard University Press.

Bowen, William J. 1949. The absorption spectra and extinction coefficients of myoglobin. *Journal of Biological Chemistry* 179: 235–45.

Boynton, Mary F. 1952. Abbott Thayer and natural history. *Osiris* 10: 542–55.

Bradbury, J. W., and S. L. Vehrencamp. 2011. *Principles of Animal Communication.* 2nd ed. Sunderland, MA: Sinauer Associates.

Bradt, Glenn W. 1932. The mammals of the Malpais, an area of black lava rock in the Tularosa Basin, New Mexico. *Journal of Mammalogy* 13: 321–28.

Brakefield, Paul M. 1987. Industrial melanism: Do we have the answers? *Trends in Ecology and Evolution* 2: 117–22.

———. 1990a. A decline of melanism in the peppered moth *Biston betularia* in the Netherlands. *Biological Journal of the Linnean Society* 39: 327–34.

———. 1990b. Genetic drift and patterns of diversity among colour-polymorphic populations of the homopteran *Philaenus spumarius* in an island archipelago. *Biological Journal of the Linnean Society* 39: 219–37.

———. 2006. Evo-devo and constraints on selection. *Trends in Ecology and Evolution* 21: 362–68.

Brakefield, P. M., and T. G. Liebert. 1990. The reliability of estimates of migration in the peppered moth *Biston betularia* and some implications for selection-migration models. *Biological Journal of the Linnean Society* 39: 335–41.

Braun, J. K., and M. A. Mares. 1989. *Neotoma micropus. Mammalian Species* 330: 1–9.

Brigham, Mark R. 1990. Prey selection by big brown bats (*Eptesicus fuscus*) and common nighthawks (*Chordeiles minor*). *The American Midland Naturalist* 124: 73–80.

Brinkley, Douglas. 2009. *The Wilderness Warrior: Theodore Roosevelt and the Crusade for America.* New York: Harper Collins.

Briscoe, A.D., and L. Chittka. 2001. The evolution of color vision in insects. *Annual Review of Entomology* 46: 471–510.

Britton, G., W.L.S. Lockley, G.A. Harriman, and T.W. Goodwin. 1977. Pigmentation of the ladybird beetle *Coccinella septempunctata* by carotenoids not of plant origin. *Nature* 266: 49–50.

Brockmann, H., and O. Völker. 1934. Der gelbe Federfarbstoff des Kanarienvogels (*Serinus canaria canaria*) und das Vorkommen von Carotinoiden bei Vögeln. *Hoppe Seylers Zeitschrift* 224: 193–215.

Brooks, Robert. 2002. Variation in female mate choice within guppy populations: Population divergence, multiple ornaments and the maintenance of polymorphism. *Genetics* 116: 343–58.

Brooks, R., and J.A. Endler. 2001. Female guppies agree to differ: Phenotypic and genetic variation in mate-choice behavior and the consequences for sexual selection. *Evolution* 55: 1644–55.

Buckland, Augustus R. 2004. "Tristram, Henry Baker (1822–1906)." In *Oxford Dictionary of National Biography*. Oxford, UK: Oxford University Press.

Bugbee, Robert E. 1942. Notes on animal occurrence and activity in the White Sands National Monument, New Mexico. *Transactions of the Kansas Academy of Science* 45: 315–21.

Bunce, M., M. Szulkin, H.R.L Lerner, I. Barnes, B. Shapiro, A. Cooper, and R.N. Holdaway. 2005. Ancient DNA provides new insights into the evolutionary history of New Zealand's extinct giant eagle. *PLoS Biology* 3: e9.

Burkhart, C.G., and C.N. Burkhart. 2005. The mole theory: Primary function of melanocytes and melanin may be antimicrobial defense and immunomodulation (not solar protection). *International Journal of Dermatology* 44: 340–42.

Cain, A.J., and P.M. Sheppard. 1954. Natural selection in *Cepaea*. *Genetics* 39: 89–116.

Caine, Nancy G. 1996. Foraging for animal prey by outdoor groups of Geoffroy's marmosets (*Callithrix geoffroyi*). *International Journal of Primatology* 17: 933–45.

Calderone, J.B., B.E. Reese, and G.H. Jacobs. 2003. Topography of photoreceptors and retinal ganglion cells in the spotted hyena (*Crocuta crocuta*). *Brain, Behavior and Evolution* 62: 182–92.

Caro, Tim. 2005. *Antipredator Defenses in Birds and Mammals*. Chicago: University of Chicago Press.

Carrasco, Marisa. 2011. Visual attention: The past 25 years. *Vision Research* 51: 1484–1525.

Carroll, Sean B. 2005. *Endless Forms Most Beautiful: The New Science of Evo Devo*. New York: W.W. Norton & Company.

———. 2006. *The Making of the Fittest: DNA and the Ultimate Forensic Record of Evolution.* New York: W. W. Norton & Company.

Carson, Rachel. 1951. *The Sea Around Us.* New York: Oxford University Press.

Carvalho, L. S., B. Knott, M. L. Berg, A. T. D. Bennett, and D. M. Hunt. 2011. Ultraviolet-sensitive vision in long-lived birds. *Proceedings of the Royal Society* B 278: 107–14.

Casadevall, A., A. L. Rosas, and J. D. Nosanchuk. 2000. Melanin and virulence in *Cryptococcus neoformans. Current Opinion in Microbiology* 3: 354–58.

Castner, J. L., and D. A. Nickle. 1995. Intraspecific color polymorphism in leaf-mimicking katydids (Orthoptera: Terrigoniidae: Pseudophyllinae: Pterochrozini). *Journal of Orthoptera Research* 4: 99–103.

Cheney, K. L., and N. J. Marshall. 2009. Mimicry in coral reef fish: How accurate is this deception in terms of color and luminance? *Behavioral Ecology* 20: 459–68.

Chiao, C-C., and T. R. Hanlon. 2001. Cuttlefish camouflage: Visual perception of size, contrast and number of white squares on artificial checkerboard substrata initiates disruptive coloration. *The Journal of Experimental Biology* 204: 2119–25.

Chun, M. M., and J. M. Wolfe. 2001. Visual attention. In *Blackwell Handbook of Perception,* edited by E. Bruce Goldstein, 272–310. Oxford, UK: Blackwell Publishing.

Clarke, Bryan. 1962a. Balanced polymorphism and the diversity of sympatric species. In *Taxonomy and Geography,* edited by David Nichols, 47–70. Cambridge, UK: Cambridge University Press, Systematics Association Publications.

———. 1962b. Natural selection in mixed populations of two polymorphic snails. *Heredity* 17: 319–45.

———. 1969. The evidence for apostatic selection. *Heredity* 24: 347–52.

Clarke, B. C., W. Arthur, D. T. Horseley, and D. T. Parkin. 1978. Genetic variation and natural selection in pulmonate molluscs. In *Systematics, Evolution and Ecology,* edited by V. Fretter and J. Peake, 220–70. Vol. 2A of *Pulmonates.* New York: Academic Press.

Clarke, C. A., and P. M. Sheppard. 1966. A local survey of the distribution of industrial melanic forms of the moth *Biston betularia* and estimates of the selective value of these in an industrial environment. *Proceedings of the Royal Society* B 165: 424–39.

Combs, Gerald F., Jr. 1998. *The Vitamins: Fundamental Aspects in Nutrition and Health.* 2nd ed. New York: Academic Press.

Common, Ian F.B. 1954. A study of the ecology of the adult bogong moth, *Agrotis infusa* (Boisd) (Lepidoptera: Noctuidae), with special reference to its behaviour during migration and aestivation. *Australian Journal of Zoology* 2: 223–63.

Conner, J.K., and D.L. Hartl 2004. *A Primer of Ecological Genetics*. Sunderland, MA: Sinauer Associates.

Connolly, R.M., A.J. Melville, and J.K. Keesing. 2002. Abundance, movement and individual identification of leafy seadragons, *Phycodurus eques* (Pisces: Syngnathidae). *Marine and Freshwater Research* 53: 777–80.

Conway Morris, Simon. 1998. *The Crucible of Creation: The Burgess Shale and the Rise of Animals*. Oxford, UK: Oxford University Press.

Cook, Laurence M. 1986. Polymorphic snails on varied backgrounds. *Biological Journal of the Linnean Society* 29: 89–99.

———. 1998. A two-stage model for *Cepaea* polymorphism. *Philosophical Transactions of the Royal Society* B 353: 1577–93.

———. 2000. Changing views on melanic moths. *Biological Journal of the Linnean Society* 69: 431–41.

———. 2003. The rise and fall of the *carbonaria* form of the peppered moth. *The Quarterly Review of Biology* 78: 399–417.

Cook, L.M., R.L.H. Dennis, and G.S. Mani. 1999. Melanic morph frequency in the peppered moth in the Manchester area. *Proceedings of the Royal Society* B 266: 293–97.

Cook, L.M., and B.S. Grant. 2000. Frequency of *insularia* during the decline in melanics in the peppered moth *Biston betularia* in Britain. *Heredity* 85: 580–585.

Cook, L.M., B.S. Grant, I.J. Saccheri, and J. Mallet. 2012. Selective bird predation on the peppered moth: The last experiment of Michael Majerus. *Biology Letters* 8: 609–12.

Cooper, Jon M. 1984. Apostatic selection on prey that match the background. *Biological Journal of the Linnean Society* 23: 221–28.

Cooper, J.M., and J.A. Allen. 1994. Selection by wild birds on artificial dimorphic prey on varied backgrounds. *Biological Journal of the Linnean Society* 51: 433–46.

Cortissoz, Royal. 1923. *American Artists*. New York: Charles Scribner's Sons.

Coston-Clements, L., L.R. Settle, D.E. Hoss and F.A. Cross. 1991. Utilization of the *Sargassum* habitat by marine invertebrates and vertebrates—A review. *NOAA Technical Memorandum NMFS-SEFSC*-296: 1–28.

Cott, Hugh B. 1957. *Adaptive Coloration in Animals*. London: Methuen.

Cotterman, T. E., D. R. Meyer, and D. D. Wickens. 1956. Discrimination reversal learning in marmosets. *Journal of Comparative and Physiological Psychology*, 49: 539–41.

Coyne, Jerry A. 1998. Not black and white. *Nature* 396: 35–36.

Cronin, T. W., N. J. Marshall, and M. F. Land. 1994. The unique visual system of the mantis shrimp. *American Scientist* 82: 356–65.

Croze, Harvey. 1970. *Searching Image in Carrion Crows: Hunting Strategy in a Predator and Some Anti-Predator Devices in Camouflaged Prey.* Berlin: Paul Parey.

Cummings, M. E., and J. C. Partridge. 2001. Visual pigments and optical habitats of surfperch (Embiotocidae) in the California kelp forest. *Journal of Comparative Physiology A* 187: 875–89.

Curio, Eberhard. 1976. *The Ethology of Predation.* Berlin: Springer-Verlag.

Cuthill, Innes C. 2006. Color perception. In *Mechanisms and Measurements*, edited by G. E. Hill and K. J. McGraw, 3–40. Vol. 1 of *Bird Coloration*. Cambridge, MA: Harvard University Press.

Cuthill, I. C., M. Stevens, J. Sheppard, T. Maddocks, C. A. Párraga, and T. S. Troscianko. 2005. Disruptive coloration and background pattern matching. *Nature* 434: 72–74.

Cuthill, I. C., M. Stevens, A. M. M. Windsor, and H. J. Walker. 2006. The effects of pattern symmetry on detection of disruptive and background-matching coloration. *Behavioral Ecology* 17: 828–32.

Cuthill, I. C., and A. Székely. 2009. Coincident disruptive coloration. *Philosophical Transactions of the Royal Society* B 364: 489–96.

Czaplicki, J. A., R. H. Porter, and H. C. Wilcoxon. 1975. Olfactory mimicry involving garter snakes and artificial models and mimics. *Behaviour* 54: 60–71.

Darwin, Charles R. 1845. *Journal of Researches into the Natural History and Geology of the Countries Visited During the Voyage Round the World of H.M.S. 'Beagle' under the Command of Captain Fitz Roy, R.N.* 2nd ed. London: John Murray.

———. 1858. Letter to Charles Lyell, June 18, 1858. Darwin Correspondence Database, www.darwinproject.ac.uk (letter #2285).

———. 1859. *On the Origin of Species by Means of Natural Selection, or the Preservation of Favoured Races in the Struggle for Life.* London: John Murray.

———. 1862. Letter to Henry W. Bates, November 20, 1862. Darwin Correspondence Database, www.darwinproject.ac.uk (letter #3816).

———. 1872. *The Origin of Species by Means of Natural Selection.* 6th ed. London: John Murray.

Darwin, C. R., and A. R. Wallace. 1858. On the tendency of species to form varieties; and on the perpetuation of varieties and species by natural means of selection. *Journal of the Proceedings of the Linnean Society of London, Zoology* 3: 45–62.

Darwin, Erasmus. 1796. *Zoonomia; or, the Laws of Organic Life*. Vol. 1. London: Joseph Johnson.

Davis, A. K., and K. L. Grayson. 2008. Spots of adult male red-spotted newts are redder and brighter than in females: Evidence for a role in mate selection? *Herpetological Journal* 18: 83–89.

Davis, Fredrick R. 2009. *Papilio dardanus:* The natural animal from the experimentalist's point of view. In *Descended from Darwin,* edited by J. Cain and M. Ruse, 221–42. Philadelphia: American Philosophical Society.

Dawkins, Marian. 1971a. Perceptual changes in chicks: Another look at the "search image" concept. *Animal Behaviour* 19: 566–74.

———. 1971b. Shifts of "attention" in chicks during feeding. *Animal Behaviour* 19: 575–82.

Dawkins, M. S., T. Halliday, and R. Dawkins, eds. 1991. *The Tinbergen Legacy*. New York: Chapman and Hall.

Dawkins, Richard. 1986. *The Blind Watchmaker*. New York: W. W. Norton & Company.

Daws, G., and M. Fujita. 1999. *Archipelago: The Islands of Indonesia*. Berkeley: University of California Press.

de Ruiter, Leendert. 1952. Some experiments on the camouflage of stick caterpillars. *Behaviour* 4: 222–32.

———.1955. Countershading in caterpillars: An analysis of its adaptive significance. *Archives Néerlandaises de Zoologie* 11: 285–341.

Dearn, John M. 1990. Color pattern polymorphism. In *Biology of Grasshoppers*, edited by R. F. Chapman and A. Joern, 517–49. New York: Wiley-IEEE Press.

Defrize, J., M. Théry, and J. Casas. 2010. Background colour matching by a crab spider in the field: A community sensory ecology perspective. *Journal of Experimental Biology* 213: 1425–35.

Degenhardt, W. G., C. W. Painter, and A. H. Price. 1996. *Amphibians and Reptiles of New Mexico*. Albuquerque: University of New Mexico Press.

del Hoyo, J., A. Elliott, and J. Sargatal, eds. 1997. *Sandgrouse to Cuckoos*. Vol. 4 of *Handbook of the Birds of the World*. Barcelona: Lynx Edicions.

Desmond, A., and J. Moore. 1991. *Darwin, the Life of a Tormented Evolutionist*. New York: W. W. Norton & Company.

Des Roches, S., J. M. Robertson, L. J. Harmon, and E. B. Rosenblum. 2011. Ecological release in White Sands lizards. *Ecology and Evolution* 1: 571–78.

Devin, S., L. Bollache, J.-N. Beisle, J.-C. Moreteau, and M.-J. Perrot-Minnot. 2004. Pigmentation polymorphism in the invasive amphipod *Dikerogammarus villosus:* Some insights into its maintenance. *Journal of Zoology* 264: 391–97.

Diamond, Jared M. 1990. Bob Dylan and moas' ghosts. *Natural History.* 99: 26–31.

Diamond, J., and A.B. Bond. 1999. *Kea, Bird of Paradox: The Evolution and Behavior of a New Zealand Parrot.* Berkeley: University of California Press.

Diamond, J., D. Eason, C. Reid, and A. Bond. 2006. Social play in kakapo (*Strigops habroptilus*) with comparisons to kea (*Nestor notabilis*) and kaka (*Nestor meridionalis*). *Behaviour* 143: 1397–1423.

Diamond, J., and P. Kociolek. 2012. Pattern and process: Natural history museum exhibits on evolution. In, *Evolution Challenges: Integrating Research and Practice in Teaching and Learning about Evolution,* edited by K.R. Rosengren, S. Brem, E.M. Evans, and G. Sinatra, 373–86. Oxford, UK: Oxford University Press.

Dice, Lee R. 1929. Description of two new pocket mice and a new woodrat from New Mexico. *Occasional Papers of the Museum of Zoology, University of Michigan* 203: 1–4.

———.1947. Effectiveness of selection by owls of deer mice (*Peromyscus maniculatus*) which contrast in color with their background. *Contributions from the Laboratory of Vertebrate Biology* 34: 1–20.

Dice, L.R., and P.M. Blossom. 1937. Studies of mammalian ecology in southwestern North America, with special attention to the colors of desert mammals. *Carnegie Institution of Washington Publication* 485: 1–129.

Dixon, Charles C. 1925. The Sargasso Sea. *The Geographical Journal* 66: 434–42.

Dixon, James R. 1967. Aspects of the biology of the lizards of the White Sands, New Mexico. *Contributions in Science* 129: 1–22.

Dobkin, David S. 1979. Functional and evolutionary relationships of vocal copying phenomena in birds. *Zeitschrift für Tierpsychologie* 50: 348–53.

Dowling, John E. 2012. *The Retina.* Cambridge, MA: Harvard University Press.

Dukas, Reuven. 2002. Behavioural and ecological consequences of limited attention. *Philosophical Transactions of the Royal Society of London* B 357: 1539–47.

Dukas, R., and A.C. Kamil. 2000. The cost of limited attention in blue jays. *Behavioral Ecology* 11: 502–6.

———. 2001. Limited attention: The constraint underlying search image. *Behavioral Ecology* 12: 192–199.

Dunbar, Nelia W. 1999. Cosmogenic ^{36}Cl-determined age of the Carrizozo lava flows, south-central New Mexico. *New Mexico Geology* 21: 25–29.

Dwyer, Philip. 2008. *Napoleon: The Path to Power.* New Haven: Yale University Press.

Dyck, Jan. 1992. Reflectance spectra of plumage areas colored by green feather pigments. *The Auk* 109: 293–301.

Eason, D.K., G.P. Elliott, D.V. Merton, P.W. Jansen, G.A. Harper, and R.J. Moorhouse. 2006. Breeding biology of kakapo (*Strigops habroptilus*) on offshore island sanctuaries, 1990–2002. *Notornis* 53: 27–36.

Edleston, Robert S. 1865. *Amphydasis betularia. The Entomologist* 2: 150.

Edmunds, Malcolm. 1972. Defensive behaviour in Ghanian praying mantids. *Zoological Journal of the Linnean Society* 51: 1–32.

———. 1974. *Defence in Animals: A Survey of Anti-predator Defences.* Burnt Mill: Longman.

———. 1976. The defensive behaviour of Ghanian praying mantids with a discussion of territoriality. *Zoological Journal of the Linnean Society* 58: 1–37.

———. 1990. The evolution of cryptic coloration. In *Insect Defenses: Adaptive Mechanisms and Strategies of Prey and Predators,* edited by D.L. Evans and J.O. Schmidt, 3–21. Albany: State University of New York Press.

Edmunds, M., and R.A. Dewhirst. 1994. The survival value of countershading with wild birds as predators. *Biological Journal of the Linnean Society* 51: 447–52.

Eisenberg, John F. 1963. The behavior of heteromyid rodents. *University of California Publications in Zoology* 69: 1–100.

Eisner, T., M. Eisner, and M. Siegler. 2005. *Secret Weapons: Defenses of Insects, Spiders, Scorpions, and Other Many-legged Creatures.* Cambridge, MA: Belknap Press.

Eizirik, E., N. Yuhki, W.E. Johnson, M. Menotti-Raymond, S.S. Hannah, and S.J. O'Brien. 2003. Molecular genetics and evolution of melanism in the cat family. *Current Biology* 13: 448–53.

Elliott, G.P., D.V. Merton, and P.W. Jansen. 2001. Intensive management of a critically endangered species: The kakapo. *Biological Conservation* 99: 121–33.

Endler, John A. 1978. A predator's view of animal color patterns. *Evolutionary Biology* 11: 319–64.

———. 1980. Natural selection and color patterns in *Poecilia reticulata. Evolution* 34: 76–91.

———. 1983. Natural and sexual selection on colour patterns in Poeciliid fishes. *Environmental Biology of Fishes* 9: 173–90.

———. 1984. Progressive background matching in moths, and a quantitative measure of crypsis. *Biological Journal of the Linnean Society* 22: 187–231.

———. 1986. *Natural Selection in the Wild.* Princeton: Princeton University Press.

———. 1987. Predation, light intensity and courtship behaviour in *Poecilia reticulata* (Pisces: Poeciliidae). *Animal Behaviour* 35: 1376–85.

———. 1990. On the measurement and classification of colour in studies of animal colour patterns. *Biological Journal of the Linnean Society* 41: 315–52.

———. 1991. Variation in the appearance of guppy color patterns to guppies and their predators under different visual conditions. *Vision Research* 31: 587–608.

———. 1993. Some general comments on the evolution and design of animal communication systems. *Philosophical Transactions of the Royal Society* B 340: 215–25.

Endler, J. A., and J. Mappes. 2004. Predator mixes and the conspicuousness of aposematic signals. *The American Naturalist* 163: 532–47.

Endler, J. A., and P. W. Mielke. 2005. Comparing entire colour patterns as birds see them. *Biological Journal of the Linnean Society* 86: 405–31.

Endler, J. A., D. A. Westcott, J. R. Madden, and T. Robson. 2005. Animal visual systems and the evolution of color patterns: Sensory processing illuminates signal evolution. *Evolution* 59: 1795–1818.

Engelmann, Theodor W. 1881. Neue Methode zur Untersuchung der Sauerstoffausscheidung Pflanzlicher und Thierischer Organismen. *Botanische Zeitung* 39: 441–48.

Ericson, P. G. P., A.-L. Jansén, U. S. Johansson, and J. Ekman. 2005. Inter-generic relationships of the crows, jays, magpies and allied groups (Aves: Corvidae) based on nucleotide sequence data. *Journal of Avian Biology* 36: 222–34.

Eterovick, P. C., J. E. C. Figueira, and J. Vasconcellos-Neto. 1997. Cryptic coloration and choice of escape microhabitats by grasshoppers (Orthoptera: Acrididae). *Biological Journal of the Linnean Society* 61: 485–99.

Evans, D. L., and J. O. Schmidt, eds. 1990. *Insect Defenses: Adaptive Mechanisms and Strategies of Prey and Predators.* Albany: State University of New York Press.

Evans, Mary Alice. 1965. Mimicry and the Darwinian heritage. *Journal of the History of Ideas* 26: 211–20.

Falmagne, Jean-Claude. 1965. Stochastic models for choice reaction time with applications to experimental results. *Journal of Mathematical Psychology* 2: 77–124.

Fernald, Russell D. 1997. The evolution of eyes. *Brain, Behavior and Evolution* 50: 253–59.

———. 2000. Evolution of eyes. *Current Opinion in Neurobiology* 10: 444–50.

Feynman, Richard. 1965. *The Character of Physical Law.* Cambridge, MA: MIT Press.

Filardi, C. E., and C. E. Smith. 2008. Social selection and geographic variation in two monarch flycatchers from the Solomon Islands. *The Condor* 110: 24–34.

Findley, J. S., A. H. Harris, D. E. Wilson, and C. Jones. 1975. *Mammals of New Mexico.* Albuquerque: University of New Mexico Press.

Fleming, Ian. 1960. *For Your Eyes Only.* London: Jonathan Cape.

Forbes, Peter. 2009. *Dazzled and Deceived: Mimicry and Camouflage.* New Haven: Yale University Press.

Ford, Edmund B. 1953. *Mendelism and Evolution.* 5th ed. London: Methuen.

———. 1948. Letter to Kettlewell, July 19, 1948. Bodleian Library Collections: University of Oxford. Cited with permission of the president and Fellows of Wolfson College, Oxford.

———. 1975. *Ecological Genetics.* London: Chapman and Hall.

———. 1979. H. B. D. Kettlewell. *Nature* 281: 166.

Forshaw, Joseph M. 2002. *Australian Parrots.* 3rd ed. Robina, Australia: Alexander Editions.

Fox, Denis L. 1962. Metabolic fractionation, storage and display of carotenoid pigments by flamingos. *Comparative Biochemistry and Physiology* 6: 1–40.

———. 1976. *Animal Biochromes and Structural Colours: Physical, Chemical, Distributional & Physiological Features of Coloured Bodies in the Animal World.* 2nd ed. Berkeley: University of California Press.

———. 1979. *Biochromy: Natural Coloration of Living Things.* Berkeley: University of California Press.

Franconeri, S. L., and D. J. Simons. 2003. Moving and looming stimuli capture attention. *Perception and Psychophysics* 65: 999–1010.

Fuertes, Louis A. 1910. A review of concealing coloration in the animal kingdom. *Science* 32: 466–69.

Fusco, Giuseppe. 2012. *Evolving Pathways: Key Themes in Evolutionary Developmental Biology* (A. Minelli, ed). Cambridge, UK: Cambridge University Press.

Fuzeau-Braesch, Suzel. 1972. Pigments and color changes. *Annual Review of Entomology* 17: 403–24.

Gehring, Walter J. 1998. *Master Control Genes in Development and Evolution: The Homeobox Story.* New Haven: Yale University Press.

———. 2005. New perspectives on eye development and the evolution of eyes and photoreceptors. *Journal of Heredity* 96: 171–84.

Gendron, Robert P. 1986. Searching for cryptic prey: Evidence for optimal search rates and the formation of search images in quail. *Animal Behaviour* 34: 898–912.

Gendron, R. P., and J. E. R. Staddon. 1983. Searching for cryptic prey: The effect of search rate. *The American Naturalist* 121: 172–86.

———. 1984. A laboratory simulation of foraging behavior: The effect of search rate on the probability of detecting prey. *The American Naturalist* 124: 407–15.

Genoways, H. H., and J. H. Brown, eds. 1993. *Biology of the Heteromyidae,* Special Publication 10. Lawrence: American Society of Mammalogists.

Gephart, Emily. 2001. Hidden talents: The camouflage paintings of Abbot [*sic*] Handerson Thayer. *Cabinet Magazine* 4: 22-25.

Getty, T., A. C. Kamil, and P. G. Real. 1987. Signal detection theory and foraging for cryptic or mimetic prey. In *Foraging Behavior,* edited by A. C. Kamil, J. R. Krebs, and H. R. Pulliam, 525–49. New York: Plenum Press.

Getty, T., and H. R. Pulliam. 1991. Random prey detection with pause-travel search. *The American Naturalist* 138: 1459–77.

———. 1993. Search and prey detection by foraging sparrows. *Ecology* 74: 734–42.

Ghiradella, Helen. 1991. Light and color on the wing: Structural colors in butterflies and moths. *Applied Optics* 30: 3492–3500.

———. 1999. Shining armor: Structural colors in insects. *Optics and Photonics News,* March 1999: 46–48.

Ghiradella, H., and M. W. Butler. 2009. Many variations on a few themes: A broader look at development of iridescent scales (and feathers). *Journal of the Royal Society Interface* 6: S243–51.

Gillis, Rick (J. E.). 1982. Substrate colour-matching cues in the cryptic grasshopper *Circotettix rabula rabula* (Rehn & Hebard). *Animal Behaviour* 30: 113–16.

Goldman, Edward A. 1932. Two new rodents from Arizona. *Proceedings of the Biological Society of Washington* 45: 89–92.

———. 1933. New mammals from Arizona, New Mexico, and Colorado. *Journal of the Washington Academy of Sciences* 23: 463–73.

Goldsmith, Timothy H. 1990. Optimization, constraint, and history in the evolution of eyes. *Quarterly Review of Biology* 65: 281–322.

Goodwin, Derek. 1986. *Crows of the World,* 2nd ed. London: British Museum (Natural History).

Goodwin, Trevor W. 1980. *Plants.* Vol. 1 of *The Biochemistry of the Carotenoids.* 2nd ed. London: Chapman and Hall.

Gorea, A., and D. Sagi. 2000. Failure to handle more than one internal representation in visual detection tasks. *Proceedings of the National Academy of Sciences* 97: 12380–4.

Gould, Stephen J. 1989. *Wonderful Life: The Burgess Shale and the Nature of History.* New York: W. W. Norton & Company.

———. 1991. Red wings in the sunset. In *Bully for Brontosaurus: Reflections in Natural History,* 209–28. New York: W. W. Norton & Company.

Grant, B. R., and P. R. Grant. 1989. *Evolutionary Dynamics of a Natural Population: The Large Cactus Finch of the Galápagos.* Chicago: University of Chicago Press.

Grant, Bruce S. 1999. Fine tuning the peppered moth paradigm. *Evolution* 53: 980–984.

———. 2004. Allelic melanism in American and British peppered moths. *Journal of Heredity* 95: 97–102.

Grant, B. S., A. D. Cook, C. A. Clarke, and D. F. Owen. 1998. Geographic and temporal variation in the incidence of melanism in peppered moth populations in America and Britain. *Journal of Heredity* 89: 465–71.

Grant, B. S., D. F. Owen, and C. A. Clarke. 1996. Parallel rise and fall of melanic peppered moths in America and Britain. *Journal of Heredity* 87: 351–57.

Grant, Peter R. 1986. *Ecology and Evolution of Darwin's Finches.* Princeton, NJ: Princeton University Press.

———. ed. 1998. *Evolution on Islands.* Oxford, UK: Oxford University Press.

Green, J., and N. M. Short. 1971. *Volcanic Landforms and Surface Features: A Photographic Atlas and Glossary.* New York: Springer-Verlag.

Greene, Erick. 1996. Effect of light quality and larval diet on morph induction in the polymorphic caterpillar *Nemoria arizonaria* (Lepidoptera: Geometridae). *Biological Journal of the Linnean Society* 58: 277–85.

Greene, Harry W. 1977. The aardwolf as hyaena mimic: An open question. *Animal Behaviour* 25: 245–46.

———. 1997. *Snakes: The Evolution of Mystery in Nature.* Berkeley: University of California Press.

Greene, H. W., and R. W. McDiarmid. 1981. Coral snake mimicry: Does it occur? *Science* 213: 1207–12.

Greene, John C. 1959. *The Death of Adam: Evolution and its Impact on Western Thought.* Ames: Iowa University Press.

Greenewalt, C. H., W. Brandt, and D. D. Friel. 1960. The iridescent colors of hummingbird feathers. *Proceedings of the American Philosophical Society* 104: 249–53.

Gregory, Richard L. 1978. *Eye and Brain: The Psychology of Seeing.* New York: McGraw-Hill.

Gutmann, J. T., and M. F. Sheridan. 1978. Geology of the Pinacate volcanic field. In *Guidebook to the Geology of Central Arizona,* vol. 2, edited by D. M. Burt and T. L. Péwé, 47–60. Tucson: Arizona Bureau of Geology and Mineral Technology.

Hacker, S. D., and L. P. Madin. 1991. Why habitat architecture and color are important to shrimps living in pelagic *Sargassum:* Use of camouflage and plant-part mimicry. *Marine Ecology Progress Series* 70: 143–55.

Halkka, O., and L. Halkka. 1990. Population genetics of the polymorphic meadow spittlebug, *Philaenus spumarius* (L.). *Evolutionary Biology* 24: 149–91.

Hamilton, William J. III. 1973. *Life's Color Code.* New York: McGraw-Hill.

Hanlon, R. T., C.-C. Chiao, L. M. Mäthger, A. Barbosa, K. C. Buresch, and C. Chubb. 2009. Cephalopod dynamic camouflage: Bridging the continuum between background matching and disruptive coloration. *Philosophical Transactions of the Royal Society* B 364: 429–37.

Hanlon, R. T., and J. B. Messenger. 1996. *Cephalopod Behaviour.* Cambridge, UK: Cambridge University Press.

Hargeby, A., J. Johansson, and J. Ahnesjö. 2004. Habitat-specific pigmentation in a freshwater isopod: Adaptive evolution over a small spatiotemporal scale. *Evolution* 58: 81–94.

Harris, A. C., and I. L. Weatherall. 1991. Geographic variation for colour in the sand-burrowing beetle *Chaerodes trachyscelides* White (Coleoptera: Tenebrionidae) on New Zealand beaches analysed using CIELAB L* values. *Biological Journal of the Linnean Society* 44: 93–104.

Harris, Moses. 1766. *The Aurelian: Or, Natural History of English Insects; Namely, Moths and Butterflies. Together with the Plants on which they Feed; and their Standard Names, as Given and Established by the Society of Aurelians.* Twickenham: Newnes.

Hart, N. S., J. C. Partridge, A. T. D. Bennett, and I. C. Cuthill. 2000. Visual pigments, cone oil droplets and ocular media in four species of estrildid finch. *Journal of Comparative Physiology* A 186: 681–94.

Hart, N. S., J. C. Partridge, I. C. Cuthill, and A. T. D. Bennett. 2000. Visual pigments, oil droplets, ocular media and cone photoreceptor distribution in two species of passerine bird: The blue tit (*Parus caeruleus* L.) and the blackbird (*Turdus merula* L.). *Journal of Comparative Physiology* A 186: 375–87.

Hartl, D. A., and A. G. Clark. 2007. *Principles of Population Genetics,* 4th ed. Sunderland, MA: Sinauer Associates.

Hassanali, A., P. G. N. Njagi, and M. O. Bashir. 2005. Chemical ecology of locusts and related acridids. *Annual Review of Entomology* 50: 223–45.

Håstad, O., J. Victorsson, and A. Ödeen 2005. Differences in color vision make passerines less conspicuous in the eyes of their predators. *Proceedings of the National Academy of Sciences* 102:6391–94.

Hedrick, Philip W. 1986. Genetic polymorphism in heterogeneous environments: A decade later. *Annual Review of Ecology and Systematics* 17: 535–66.

———. 2006. Genetic polymorphism in heterogeneous environments: The age of genomics. *Annual Review of Ecology and Systematics* 37: 67–93.

Heiling, A. M., K. Cheng, and M. E. Herberstein. 2006. Picking the right spot: Crab spiders position themselves on flowers to maximize prey attraction. *Behaviour* 143: 957–68.

Heiling, A., L. Chittka, K. Cheng, and M. E. Herberstein. 2005. Colouration in crab spiders: Substrate choice and prey attraction. *Journal of Experimental Biology* 208: 1785–92.

Heim de Balsac, Henri. 1936. Biogéographie des mammifères et des oiseaux de l'Afrique du Nord. *Bulletin Biologique de France et de Belgique, Supplément* 21: 1–446.

Heinsohn, R., S. Legge, and J. A. Endler. 2005. Extreme reversed sexual dichromatism in a bird without sex role reversal. *Science* 309: 617–19.

Henry, John. 2008. *The Scientific Revolution and the Origins of Modern Science.* 3rd ed. New York: Palgrave Macmillan.

Higgins, Peter J., ed. 1999. *Parrots to Dollarbird.* Vol. 4 of *Handbook of Australian, New Zealand and Antarctic Birds.* South Melbourne, Australia: Oxford University Press.

Hill, Helene Z. 1992. The function of melanin or six blind people examine an elephant. *Bioessays* 14: 49–56.

Himmelman, John. 2011. *Cricket Radio: Tuning in the Night-singing Insects.* Cambridge, MA: Belknap Press.

Hines, H. M., B. A. Counterman, R. Papa, P. A. de Moura, M. Z. Cardoso, M. Linares, J. Mallet, R. D. Reed, C. D. Jiggins, M. R. Kronforst, and W. O. McMillan. 2011. Wing patterning gene redefines the mimetic history of *Heliconius* butterflies. *Proceedings of the National Academy of Sciences* 108: 19666–71.

Hinton, Howard E. 1976. Possible significance of the red patches of the female crab-spider, *Misumena vatia. Journal of Zoology* 180: 35–39.

Hoekstra, H. E., K. E. Drumm, and M. W. Nachman. 2004. Ecological genetics of adaptive color polymorphism in pocket mice: Geographic variation in selected and neutral genes. *Evolution* 58: 1329–41.

Hoekstra, H. E., J. G. Krenz, and M. W. Nachman. 2005. Local adaptation in the rock pocket mouse (*Chaetodipus intermedius*): Natural selection and phylogenetic history of populations. *Heredity* 94: 217–28.

Hoekstra, H. E., and M. W. Nachman. 2003. Different genes underlie adaptive melanism in different populations of rock pocket mice. *Molecular Ecology* 12: 1185–94.

———. 2005. Coat color variation in rock pocket mice (*Chaetodipus intermedius*): From genotype to phenotype. In *Mammalian Diversification: From Chromosomes to Phylogeography*, edited by E. A. Lacey and P. Myers, 79–100. Berkeley: University of California Press.

Hoffman, Donald D. 1998. *Visual Intelligence: How We Create What We See.* New York: W. W. Norton & Company.

Hoffman, E. A., and M. S. Blouin. 2000. A review of colour and pattern polymorphisms in anurans. *Biological Journal of the Linnean Society* 70: 633–65.

Hoffmeister, Donald F. 1986. *Mammals of Arizona.* Tucson: University of Arizona Press.

Hölldobler, B., and E. O. Wilson. 1990. *The Ants.* Cambridge, MA: Harvard University Press.

Holling, Crawford S. 1965. The functional response of predators to prey density and its role in mimicry and population regulation. *Memoirs of the Entomological Society of Canada* 45: 1–60.

Hooper, Emmet T. 1941. Mammals of the lava fields and adjoining areas in Valencia County, New Mexico. *Miscellaneous Publications of the Museum of Zoology, University of Michigan* 51: 1–53.

Hooper, Judith. 2002. *Of Moths and Men: An Evolutionary Tale: The Untold Story of Science and the Peppered Moth.* New York: W. W. Norton & Company.

Horecker, Bernard L. 1943. The absorption spectra of hemoglobin and its derivatives in the visible and near infra-red regions. *The Journal of Biological Chemistry* 148: 173–83.

Horowitz, Alexandra. 2009. *Inside of a Dog: What Dogs See, Smell, and Know.* New York: Scribners.

Horstmann, Gernot. 2002. Evidence for attentional capture by a surprising color singleton in visual search. *Psychological Science* 13: 499–505.

Houde, Anne E. 1997. *Sex, Color, and Mate Choice in Guppies.* Princeton: Princeton University Press.

Hughes, K. A., L. Du, F. H. Rodd, and D. N. Reznick. 1999. Familiarity leads to female mate preference for novel males in the guppy, *Poecilia reticulata. Animal Behaviour* 58: 907–16.

Humphreys, G. W., and V. Bruce. 1989. *Visual Cognition: Computational, Experimental, and Neuropsychological Perspectives.* Hillsdale: Lawrence Erlbaum Associates.

Huxley, Julian. 1974. *Evolution: The Modern Synthesis.* London: Allen & Unwin.

Israel, Jonathan I. 2001. *Radical Enlightenment: Philosophy and the Making of Modernity 1650–1750.* Oxford, UK: Oxford University Press.

Itti, L., and C. Koch. 2000. A saliency-based search mechanism for overt and covert shifts of visual attention. *Vision Research* 40: 1489–1506.

———. 2001. Computational modeling of visual attention. *Nature Reviews Neuroscience* 2: 194–203.

Jackson, Michael H. 1993. *Galápagos, A Natural History.* Calgary, Canada: University of Calgary Press.

Jacobs, G. H., M. Crognale, and J. Fenwick. 1987. Cone pigment in the great horned owl. *Condor* 89: 434–36.

Jacobs, G. H., J. F. Deegan II, and J. Neitz. 1998. Photopigment basis for dichromatic color vision in cows, goats, and sheep. *Visual Neuroscience* 15: 581–84.

Jacobs, G. H., and J. Neitz. 1986. Spectral mechanisms and color vision in the tree shrew (*Tupaia belangeri*). *Vision Research* 26: 291–98.

Jacobson, E. S., E. Hove, and H. S. Emery. 1995. Antioxidant function of melanin in black fungi. *Infection and Immunity* 63: 4944–45.

Jaeger, Edmund C. 1957. *The North American Deserts.* Stanford, CA.: Stanford University Press.

al-Jāḥiẓ 1969. *The Life and Works of Jāḥiẓ: Translations of Selected Texts.* Translated by Charles Pellat. Translated from the French by D. M. Hawke. Berkeley: University of California Press.

James, William. 1890. *The Principles of Psychology.* New York: Henry Holt and Company.

Johnsen, Sönke. 2001. Hidden in plain sight: The ecology and physiology of organismal transparency. *Biological Bulletin* 201: 301–18.

Johnson, Kristin. 2007. Natural history as stamp collecting: A brief history. *Archives of Natural History* 34: 244–58.

Jones, C. D., and D. Osorio. 2004. Discrimination of oriented visual textures by poultry chicks. *Vision Research* 44: 83–89.

Jørgensen, Jørgen M. 1991. Regeneration of lateral line and inner ear vestibular cells. *CIBA Foundation Symposia* 160: 151–70.

———. 1992. The electrosensory cells of the ampullary organ of the transparent catfish (*Kryptopterus bicirrhus*). *Acta Zoologica* 73: 79–83.

Joron, M., C. D. Jiggins, A. Papanicolaou, and W. O. McMillan. 2006. *Heliconius* wing patterns: An evo-devo model for understanding phenotypic diversity. *Heredity* 97: 157–67.

Judd, Sylvester D. 1899. The efficiency of some protective adaptations in securing insects from birds. *The American Naturalist* 33: 461–84.

———. 1905. The grouse and wild turkeys of the United States, and their economic value. *U.S. Department of Agriculture Biological Survey Bulletin* 24: 1–55.

Kamil, A.C., and A.B. Bond 2006. Selective attention, priming, and foraging behavior. In *Comparative Cognition: Experimental Explorations of Animal Intelligence,* edited by E.A. Wasserman and T.R. Zentall, 106–26. Oxford, UK: Oxford University Press.

Kamil, A.C., T.B. Jones, A. Pietrewicz, and J.E. Mauldin. 1977. Positive transfer from successive reversal training to learning set in blue jays (*Cyanocitta cristata*). *Journal of Comparative and Physiological Psychology* 91: 79–86.

Kaufman, Donald W. 1974. Adaptive coloration in *Peromyscus polionotus:* Experimental selection by owls. *Journal of Mammalogy* 55: 271–83.

Kelley, J.L., J.A. Graves, and A.E. Magurran. 1999. Familiarity breeds contempt in guppies. *Nature* 401: 661–62.

Kennedy, Gilbert Y. 1975. Porphyrins in invertebrates. *Annals of the New York Academy of Sciences* 244: 662–73.

Kettlewell, H. Bernard D. 1955. Selection experiments on industrial melanism in the Lepidoptera. *Heredity* 9: 323–42.

———. 1956. Further selection experiments on industrial melanism in the Lepidoptera. *Heredity* 10: 287–301.

———. 1958. A survey of the frequencies of *Biston betularia* (L.) (Lep.) and its melanic forms in Great Britain. *Heredity* 12: 51–72.

———. 1973. *The Evolution of Melanism.* Oxford, UK: Clarendon Press.

Kettlewell, H.B.D., and D.L.T. Conn. 1977. Further background-choice experiments on cryptic Lepidoptera. *Journal of Zoology* 181: 371–76.

Kikuchi, D.W., and D.W. Pfennig. 2010. Predator cognition permits imperfect coral snake mimicry. *American Naturalist* 176: 830–34.

Kiltie, Richard A. 1988. Countershading: Universally deceptive or deceptively universal? *Trends in Ecology & Evolution* 3: 21–23.

Kingsland, Sharon. 1978. Abbott Thayer and the protective coloration debate. *Journal of the History of Biology* 11: 223–44.

Kinoshita, Shuichi. 2008. *Structural Colors in the Realm of Nature.* Hackensack: World Scientific Publishing Company.

Kipling, Rudyard. 1914. *Just So Stories.* Garden City, NY: Doubleday, Page & Co.

Kirby, W., and W. Spence. 1843. *An Introduction to Entomology: Or, Elements of the Natural History of Insects.* Vol. 2. 6th ed. London: Longman, Brown, Green & Longmans. (Reprinted by Elibron Classics, 2005).

Kollias, Nicholas. 1995. The spectroscopy of human melanin pigmentation. In *Melanin: Its Role in Human Photoprotection,* edited by L. Zeise, M.R.

Chedekel, and T.B. Fitzpatrick, 31–38. Overland Park, KS: Valdenmar Publishing Company.

Kotler, B.P., and J.S. Brown. 1988. Environmental heterogeneity and the coexistence of desert rodents. *Annual Review of Ecology and Systematics* 19: 281–307.

Koyama, Yasushi. 1991. Structures and functions of carotenoids in photosynthetic systems. *Journal of Photochemistry and Photobiology B: Biology* 9: 265–80.

Kozmik, Z., J. Ruzickova, K. Jonasova, Y. Matsumoto, P. Vopalensky, I. Kozmikova, H. Strnad, S. Kawamura, J. Piatigorsky, V. Paces, and C. Vlcek. 2008. Assembly of the cnidarian camera-type eye from vertebrate-like components. *Proceedings of the National Academy of Sciences* 105: 8989–93.

Kram, Y.A., S. Mantey, and J.C. Corbo. 2010. Avian cone photoreceptors tile the retina as five independent, self-organizing mosaics. *PLoS ONE* 5: e8992.

Krauss, Hermann. 1890. Erklärung der Orthopteren-Tafeln J.C. Savigny's in der "Description de l'Égypte." Aus der Literatur zusammengestellt und mit Bemerkungen versehen. *Verhandlungen der Zoologisch-Botanischen Gesellschaft in Wien* 40: 227–72.

Krebs, John R. 1973. Behavioral aspects of predation. In *Perspectives in Ethology,* vol. 1, edited by P.P.G. Bateson and P.H. Klopfer, 73–111. New York: Plenum Press.

Kristjánsson, Á., and G. Campana. 2010. Where perception meets memory: A review of repetition priming in visual search tasks. *Attention, Perception, & Psychophysics* 72: 5–18.

Kronforst, M.R., and L.E. Gilbert. 2008. The population genetics of mimetic diversity in *Heliconius* butterflies. *Proceedings of the Royal Society of London* B 275: 493–500.

Kunte, Krushnamegh. 2009. The diversity and evolution of Batesian mimicry in *Papilio* swallowtail butterflies. *Evolution* 63: 2707–16.

Land, Michael F. 1997. Visual acuity in insects. *Annual Review of Entomology* 42: 147–77.

———. 2003. The spatial resolution of the pinhole eyes of giant clams (*Tridacna maxima*). *Proceedings of the Royal Society* B 270: 185–88.

Land, M.F., and R.D. Fernald. 1992. The evolution of eyes. *Annual Review of Neuroscience* 15: 1–29.

Land, M.F., and D.E. Nilsson. 2002. *Animal Eyes.* Oxford, UK: Oxford University Press.

Landrum, J.T., D. Callejas, and F. Alvarez-Calderon. 2010. Specific accumulation of lutein within the epidermis of butterfly larvae. In *Carotenoids:*

Physical, Chemical and Biological Functions and Properties, edited by John T. Landrum, 525–35. New York: CRC Press.

Lane, Nick. 2009. *Life Ascending: The Ten Great Inventions of Evolution*. New York: W.W. Norton & Company.

Langley, Cynthia M. 1996. Search images: Selective attention to specific visual features of prey. *Journal of Experimental Psychology: Animal Behavior Processes* 22: 152–63.

Langley, C.M., D.A. Riley, A.B. Bond, and N. Goel. 1995. Visual search for natural grains in pigeons: Search images and selective attention. *Journal of Experimental Psychology: Animal Behavior Processes* 22: 139–51.

Langstroth, L., and L. Langstroth. 2000. *A Living Bay: The Underwater World of Monterey Bay*. Berkeley: University of California Press and Monterey Bay Aquarium.

Large, M.C.J., S. Wickham, J. Hayes, and L. Poladian. 2007. Insights from nature: Optical biomimetics. *Physica B* 394: 229–32.

Lees, David R. 1979. Obituary of Bernard Kettlewell. *Entomologist's Monthly Magazine* 114: 251–55.

Lees, D.R., and E.R. Creed. 1977. The genetics of the *insularia* forms of the peppered moth, *Biston betularia*. *Heredity* 39: 67–73.

Lees, D.R., E.R. Creed, and J.G. Duckett. 1973. Atmospheric pollution and industrial melanism. *Heredity* 30: 227–32.

Lefébvre de Cérisy, Alexandre Louis. 1836. New group of Orthoptera, Family of Mantides. *The Entomological Magazine* 4: 66–76.

Leimar, Olof. 2009. Environmental and genetic cues in the evolution of phenotypic polymorphism. *Evolutionary Ecology* 23: 125–35.

Levine, J.S., and E.F. MacNichol. 1982. Color vision in fishes. *Scientific American* 246: 140–49.

Lewis, Thomas H. 1949. Dark coloration in the reptiles of the Tularosa Malpais, New Mexico. *Copeia* 1949: 181–84.

———. 1951. Dark coloration in the reptiles of the Malpais of the Mexican border. *Copeia* 1951: 311–12.

Li, Q., K.-Q. Gao, Q. Meng, J.A. Clarke, M.D. Shawkey, L. D'Alba, R. Pei, M. Ellison, M.A. Norell, and J. Vinther. 2012. Reconstruction of *Microraptor* and the evolution of iridescent plumage. *Science* 335: 1215–19.

Li, Q., K.-Q. Gao, J. Vinther, M.D. Shawkey, J.A. Clarke, L. D'Alba, Q. Meng, D.E.G. Briggs, and R.O. Prum. 2012. Plumage color patterns of an extinct dinosaur. *Science* 327: 1369–72.

Liebert, T.G., and P.M. Brakefield. 1987. Behavioural studies on the peppered moth *Biston betularia* and a discussion of the role of pollution and lichens in industrial melanism. *Biological Journal of the Linnean Society* 31: 129–50.

Lindberg, J. D., and M. S. Smith. 1973. Reflectance spectra of gypsum sand from the White Sands National Monument and basalt from a nearby lava flow. *American Mineralogist* 58: 1062–64.

Linnaeus, Carl. 1758. *Systema Naturæ.* Holmiæ: Impensis Direct. Laurentii salvii.

Llandres, A. L., F. M. Gawryszewski, A. M. Heiling, and M. E. Herberstein. 2011. The effect of colour variation in predators on the behaviour of pollinators: Australian crab spiders and native bees. *Ecological Entomology* 36: 72–81.

Losos, J. B., and R. E. Ricklefs. 2009. Adaptation and diversification on islands. *Nature* 457: 830–36.

Lythgoe, John N. 1979. *The Ecology of Vision.* Oxford: Clarendon Press.

Mace, R. D., and C. J. Jonkel. 1986. Local food habits of the grizzly bear in Montana. *International Conference on Bear Research and Management* 6: 105–10.

Macêdo, R. H., and M. A. Mares. 1988. *Neotoma albigula. Mammalian Species* 310: 1–7.

Mack, A., and I. Rock. 2000. *Inattentional Blindness.* Cambridge, MA: MIT Press.

Mackintosh, Nicholas J. 1974. *The Psychology of Animal Learning.* New York: Academic Press.

Madge, S., and H. Burn. 1994. *Crows and Jays: A Guide to the Crows, Jays and Magpies of the World.* New York: Houghton Mifflin.

Magurran, Anne E. 2005. *Evolutionary Ecology: The Trinidadian Guppy.* Oxford, UK: Oxford University Press.

Maia, R., J. V. O. Caetano, S. N. Báo, and R. H. Macedo. 2009. Iridescent structural colour production in male blue-black grassquit feather barbules: The role of keratin and melanin. *Journal of the Royal Society Interface* 6: S203–11.

Majerus, Michael E. N. 1998. *Melanism: Evolution in Action.* Oxford: Oxford University Press.

———. 2002. *Moths.* London: Harper Collins Publishers.

———. 2005. The peppered moth: Decline of a Darwinian disciple. In *Insect Evolutionary Ecology: Proceedings of the Royal Entomological Society's 22nd Symposium,* edited by M. D. E. Fellowes, G. J. Holloway, and J. Rolff, 367–92. Cambridge, MA: CABI Publishing.

———. 2007. The Peppered Moth: The Proof of Darwinian Evolution. Talk delivered at the 11th Congress of the European Society for Evolutionary

Biology, August 20–25, 2007, Uppsala, Sweden. Talk accessed February 6, 2012 at www.gen.cam.ac.uk/research/personal/majerus/majerus.html.

———. 2009. Industrial melanism in the peppered moth, *Biston betularia:* An excellent teaching example of Darwinian evolution in action. *Evolution: Education and Outreach* 2: 63–74.

Majerus, M.E.N., and N.I. Mundy. 2003. Mammalian melanism: Natural selection in black and white. *TRENDS in Genetics* 19: 585–88.

Maljkovic, V., and P. Martini. 2005. Implicit short-term memory and event frequency effects in visual search. *Vision Research* 45: 2831–46.

Mallet, J., and M. Joron. 1999. Evolution of diversity in warning color and mimicry: Polymorphisms, shifting balance, and speciation. *Annual Review of Ecology and Systematics* 30: 201–33.

Mani, Gopalakrishnan S. 1990. Theoretical models of melanism in *Biston betularia*—a review. *Biological Journal of the Linnean Society* 39: 355–371.

Mani, G.S., and M.E.N. Majerus. 1993. Peppered moth revisited: Analysis of recent decreases in melanic frequency and predictions for the future. *Biological Journal of the Linnean Society* 48: 157–65.

Mason, R.T., and D. Crews. 1986. Pheromone mimicry in garter snakes. In *Chemical Signals in Vertebrates 4: Ecology, Evolution, and Comparative Biology,* edited by D. Duvall, D. Müller-Schwarze, and R.M. Silverstein, 279–83. New York: Plenum Press.

Mäthger, L.M., C.-C. Chiao, A. Barbosa, and R.T. Hanlon. 2008. Color matching on natural substrates in cuttlefish, *Sepia officinalis. Journal of Comparative Physiology* A 194: 577–85.

Matthews, R.W., and J.R. Matthews. 1978. *Insect Behavior.* New York: John Wiley & Sons.

Maynard Smith, John. 1962. Disruptive selection, polymorphism and sympatric speciation. *Nature* 195: 60–62.

———. 1970. Genetic polymorphism in a varied environment. *The American Naturalist* 104: 487–90.

———. 1993. *The Theory of Evolution.* Cambridge, UK: Cambridge University Press.

Mayr, Ernst. 1942. *Systematics and the Origin of Species from the Viewpoint of a Zoologist.* New York: Columbia University Press.

———. 1955. Karl Jordan's contributions to current concepts in systematics and evolution. *Transactions of the Royal Entomological Society of London* 107: 45–66.

———. 1966. *Animals Species and Evolution.* Cambridge, MA: Belknap Press.

———. 1967. The challenge of island faunas. *Australian Natural History* 15: 369–74.

Mayr, E., and J. Diamond. 2001. *The Birds of Northern Melanesia.* Oxford: Oxford University Press.

McGraw, Kevin J. 2006. Mechanics of uncommon colors: Pterins, porphyrins, and psittacofulvins. In *Mechanisms and Measurements,* edited by G.E. Hill and K.J. McGraw, 345–98. Vol. 1 of *Bird Coloration.* Cambridge, MA: Harvard University Press.

McIlwain, James T. 1996. *An Introduction to the Biology of Vision.* Cambridge, UK: Cambridge University Press.

McNair, James N. 1980. A stochastic foraging model with predator training effects: I. Functional response, switching, and run lengths. *Theorectical Population Biology* 17:141–66.

McRobie, H., A. Thomas, and J. Kelly. 2009. The genetic basis of melanism in the grey squirrel (*Sciurus carolinensis*). *Journal of Heredity* 100: 709–14.

Meinertzhagen, Richard. 1934. The relation between plumage and environment, with special reference to the Outer Hebrides. *Ibis* 76: 52–61.

Menzel, E.W., Jr., and C. Juno. 1982. Marmosets (*Saguinus fusicollis*): Are learning sets learned? *Science* 217: 750–52.

Menzel, E.W., Jr., and C.R. Menzel. 1979. Cognitive, developmental and social aspects of responsiveness to novel objects in a family group of marmosets (*Saguinus fuscicollis*). *Behaviour* 70: 251–79

Merilaita, Sami. 2003. Visual background complexity facilitates the evolution of camouflage. *Evolution* 57: 1248–54.

Merriam, C. Hart. 1890. Results of a biological survey of the San Francisco Mountain region and desert of the Little Colorado in Arizona. *North American Fauna* 3: 1–128.

———. 1918. Review of the grizzly and big brown bears of North America (genus *Ursus*) with a description of a new genus, *Vetularctos. North American Fauna* 41: 1–136.

Merrill, Elmer D. 1916. Osbeck's *Dagbok Öfwer en Ostindsk Resa. American Journal of Botany* 3: 571–88.

Messenger, John B. 2001. Cephalopod chromatophores: Neurobiology and natural history. *Biological Reviews* 76: 473–528.

Miles, R.C., and D.R. Meyer. 1956. Learning sets in marmosets. *Journal of Comparative and Physiological Psychology* 49: 219–22.

Milgrom, Lionel R. 1997. *The Colours of Life: An Introduction to the Chemistry of Porphyrins and Related Compounds.* New York: Oxford University Press.

Miller, E.S., G. Mackinney, and F.P. Zscheile, Jr. 1935. Absorption spectra of alpha and beta carotenes and lycopene. *Plant Physiology* 10: 375–81.

Millière, Pierre. 1864. *Iconographie et Description de Chenilles et Lépidoptères Inédits.* Vol. 2. Paris: F. Savy.

Moment, Gairdner B. 1962. Reflexive selection: A possible answer to an old puzzle. *Science* 136: 262–63.

Monk, R. R., and J. K. Jones. 1996. Perognathus flavescens. *Mammalian Species* 525: 1–4.

Morrone, M. C., V. Denti, and D. Spinelli. 2002. Color and luminance contrasts attract independent attention. *Current Biology* 12: 1134–37.

Morse, Douglass H. 2007. *Predator Upon a Flower: Life History and Fitness in a Crab Spider.* Cambridge, MA: Harvard University Press.

Mosley, Stephen. 2008. *The Chimney of the World: A History of Smoke Pollution in Victorian and Edwardian Manchester.* New York: Routledge.

Mundy, Nick I. 2005. A window on the genetics of evolution: *MC1R* and plumage colouration in birds. *Proceedings of the Royal Society* B 272: 1633–40.

Murdoch, W. W., and A. Oaten. 1974. Predation and population stability. *Advances in Ecological Research* 9: 1–131.

Murton, Ronald K. 1971. The significance of a specific search image in the feeding behaviour of the wood-pigeon. *Behaviour* 40: 10–42.

Nabours, R. K., I. Larson, and N. Hartwig. 1933. Inheritance of color patterns in the grouse locust *Acrydium arenosum* Burmeister (Tettigidae). *Genetics* 18: 159–71.

Nachman, Michael W. 2005. The genetic basis of adaptation: lessons from concealing coloration in pocket mice. *Genetica* 123: 125–36.

Nachman, M. W., H. E. Hoekstra, and S. L. D'Agostino. 2003. The genetic basis of adaptive melanism in pocket mice. *Proceedings of the National Academy of Sciences* 100: 5268–73.

Naskrecki, Piotr. 2005. *The Smaller Majority.* Cambridge, MA: Belknap Press.

Needham, Arthur E. 1974. *The Significance of Zoochromes.* Vol. 3 of *Zoophysiology and Ecology.* New York: Springer-Verlag.

Neitz, J., T. Geist, and G. H. Jacobs. 1989. Color vision in the dog. *Visual Neuroscience* 3: 119–25.

Nemerov, Alexander. 1997. Vanishing Americans: Abbott Thayer, Theodore Roosevelt, and the attraction of camouflage. *American Art* 11: 50–81.

Neumeyer, Christa. 1992. Tetrachromatic color vision in goldfish: Evidence from color mixture experiments. *Journal of Comparative Physiology* A 171: 639–49.

Nijhout, H. Frederik. 1991. *The Development and Evolution of Butterfly Wing Patterns.* Washington, D.C.: Smithsonian Institution Scholarly Press.

Nikolai, J. C., and D. M. Bramble. 1983. Morphological structure and function in desert heteromyid rodents. *Great Basin Naturalist Memoirs* 7: 44–64.

Nilsson, D.-E., and S. Pelger. 1994. A pessimistic estimate of the time required for an eye to evolve. *Proceedings of the Royal Society* B 256: 53–58.

Noor, M. A. F., R. S. Parnell, and B. S. Grant. 2008. A reversible color polyphenism in American peppered moth (*Biston betularia cognataria*) caterpillars. *PLoS ONE* 3: e3124.

Norris, Kenneth S. 1958. The evolution and systematics of the iguanid genus *Uma* and its relation to the evolution of other North American desert reptiles. *Bulletin of the American Museum of Natural History* 114: 247–326.

Norris, K. S., and C. H. Lowe. 1964. An analysis of background color-matching in amphibians and reptiles. *Ecology* 45: 565–80.

Nosil, Patrik. 2007. Divergent host plant adaptation and reproductive isolation between ecotypes of *Timema cristinae* walking sticks. *American Naturalist* 169: 151–162.

Nosil, P., C. P. Sandoval, and B. J. Crespi. 2006. The evolution of host preference in allopatric vs. parapatric populations of *Timema cristinae* walking-sticks. *Journal of Evolutionary Biology* 19: 929–42.

Ödeen, A., and O. Håstad. 2003. Complex distribution of avian color vision systems revealed by sequencing the SWS1 opsin from total DNA. *Molecular Biology and Evolution* 20:855–61.

Olendorf, R., F. H. Rodd, D. Punzalan, A. E. Houde, C. Hurt, D. N. Reznick, and K. A. Hughes. 2006. Frequency-dependent survival in natural guppy populations. *Nature* 441: 633–36.

Osbeck, Pehr. 1771. *A Voyage to China and the East Indies.* Translated by John R. Forster. London: Benjamin White.

Osorio, D., A. Miklósi, and Z. Gonda. 1999. Visual ecology and perception of coloration patterns by domestic chicks. *Evolutionary Ecology* 13: 673–89.

Osorio, D., and M. V. Srinivasan. 1991. Camouflage by edge enhancement in animal coloration patterns and its implications for visual mechanisms. *Proceedings of the Royal Society* B 244: 81–85.

Osorio, D., and M. Vorobyev. 1997. *Sepia* tones, stomatopod signals and the uses of color. *TRENDS in Ecology and Evolution* 12: 167–68.

———. 2008. A review of the evolution of animal colour vision and visual communication signals. *Vision Research* 48: 2042–51.

Otte, D., and P. Brock. 2003. *Phasmida Species File: A Catalog of the Stick and Leaf Insects of the World.* Philadelphia: The Orthopterists' Society.

Owen, Denis. 1980. *Camouflage and Mimicry.* Chicago: University of Chicago Press.

Owen, D. F., and D. Whiteley. 1986. Reflexive selection: Moment's hypothesis resurrected. *Oikos* 47: 117–20.

———. 1988. The beach clams of Thessaloniki: Reflexive or apostatic selection? *Oikos* 51: 253–55.

———. 1989. Evidence that reflexive polymorphisms are maintained by visual selection by predators. *Oikos* 55: 130–33.

Oxford, Geoff S. 2009. An exuberant, undescribed colour polymorphism in *Theridion californicum* (Araneae, Theridiidae): Implications for a theridiid pattern ground plan and the convergent evolution of visible morphs. *Biological Journal of the Linnean Society* 96: 23–34.

Oxford, G.S. and R.G. Gillespie. 1998. Evolution and ecology of spider coloration. *Annual Review of Entomology* 43: 619–43.

Ożgo, Małgorzata. 2011. Rapid evolution in unstable habitats: A success story of the polymorphic land snail *Cepaea nemoralis* (Gastropoda: Pulmonata). *Biological Journal of the Linnean Society* 102: 251–62.

Palma, A.T., and R.S. Steneck. 2001. Does variable coloration in juvenile marine crabs reduce risk of visual predation? *Ecology* 82: 2961–67.

Parasuraman, R., and D.R. Davies. 1984. *Varieties of Attention.* New York: Academic Press.

Pashler, Harold E. 1998. *The Psychology of Attention.* Cambridge, MA: MIT Press.

Passos, F. de C., and A. Keuroghlian. 1999. Foraging behavior and microhabitats used by black lion tamarins, *Leontopithecus chrysopygus* (Mikan) (Primates, Callitrichidae). *Revista Brasileira de Zoolgia* 16, suppl.2: 219–22.

Paterson, J.R., D.C. García-Bellido, M.S.Y. Lee, G.A. Brock, J.B. Jago, and G.D. Edgecombe. 2011. Acute vision in the giant Cambrian predator *Anomalocaris* and the origin of compound eyes. *Nature* 480: 237–40.

Payne, Robert B. 1973. Vocal mimicry of the paradise whydahs (*Vidua*) and response of female whydahs to the songs of their hosts (*Ptylia*) and their mimics. *Animal Behaviour* 21: 762–71.

———. 1976. Song mimicry and species relationships among the West African pale-winged indigobirds. *The Auk* 93: 25–38.

Payne, R.B., L.L. Payne, and J.L. Woods. 1998. Song learning in brood-parasitic indigobirds *Vidua chalybeate:* song mimicry of the host species. *Animal Behaviour* 55: 1537–53.

Peichl, Leo. 2005. Diversity of mammalian photoreceptor properties: Adaptations to habitat and lifestyle? *Anatomical Record* A 287A: 1001–12.

Pellissier, L., J. Wassef, J. Bilat, G. Brazzola, P. Buri, C. Colliard, B. Fournier, J. Hausser, G. Yannic, and N. Perrin. 2011. Adaptive colour polymorhism of *Aerida ungarica* H. (Orthoptera: Acrididae) in a spatially heterogeneous environment. *Acta Oecologica* 37: 93–8.

Pianka, Eric R. 1986. *Ecology and Natural History of Desert Lizards.* Princeton, NJ: Princeton University Press.

Pianka, E.R., and L.J. Vitt. 2003. *Lizards: Windows to the Evolution of Diversity.* Berkeley: University of California Press.

Piersma, T., and J.A. van Gils. 2011. *The Flexible Phenotype: A Body-Centred Integration of Ecology, Physiology, and Behavior.* Oxford: Oxford University Press.

Pietrewicz, A.T., and A.C. Kamil. 1977. Visual detection of cryptic prey by blue jays (*Cyanocitta cristata*). *Science* 195: 580–82.

———. 1979. Search image formation in the blue jay (*Cyanocitta cristata*). *Science* 204: 1332–33.

Pietsch, T.W., and D.B. Grobecker. 1987. *Frogfishes of the World: Systematics, Zoogeography, and Behavioral Ecology.* Stanford: Stanford University Press.

Plinius Secundus, Gaius. 1601. *The Historie of the World. Commonly Called: The Natural Historie of C. Plinius Scundus.* Translated by Philemon Holland. London: Adam Islip.

Popham, Edward J. 1941. The variation in the colour of certain species of *Arctorcorisa* (Hemiptera, Corixidae) and its significance. *Proceedings of the Zoological Society London* A 111: 135–72.

Portman, Adolf. 1959. *Animal Camouflage.* Ann Arbor: University of Michigan Press.

Pough, F. Harvey. 1988. Mimicry of vertebrates: Are the rules different? *The American Naturalist* 131: S67–S102.

Poulton, Edward B. 1890. *The Colours of Animals: Their Meaning and Use, Especially Considered in the Case of Insects.* New York: D. Appleton and Company.

Prakash, I., and P.K. Ghosh, eds. 1975. *Rodents in Desert Environments.* The Hague: Dr. W. Junk.

Pratt, J., P.V. Radulescu, R.M. Guo, and R.A. Abrams. 2010. It's alive! Animate motion captures visual attention. *Psychological Science* 21: 1724–30.

Prota, Giuseppe. 1992. *Melanins and Melanogenesis.* New York: Academic Press.

Protas, M.E., and N.H. Patel. 2008. Evolution of coloration patterns. *Annual Review of Cellular and Developmental Biology* 24: 425–46.

Provine, William B. 1986. *Sewall Wright and Evolutionary Biology.* Chicago: University of Chicago Press.

———. 2004. Ernst Mayr: Genetics and speciation. *Genetics* 167: 1041–46.

Prum, Richard O. 2006. Anatomy, physics, and evolution of structural colors. In *Mechanisms and Measurements,* edited by G.E. Hill and K.J. McGraw, 295–353. Vol. 1 of *Bird Coloration.* Cambridge, MA: Harvard University Press.

Punzalan, D., F.H. Rodd, and K.A. Hughes. 2005. Perceptual processes and the maintenance of polymorphism through frequency-dependent predation. *Evolutionary Ecology* 19: 303–20.

Raboy, B.E., and J.M. Dietz. 2004. Diet, foraging, and use of space in wild golden-headed lion tamarins. *American Journal of Primatology* 63: 1–15.

Raby, Peter. 2001. *Alfred Russel Wallace: A Life*. Princeton: Princeton University Press.

Randall, Jan A. 1993. Behavioural adaptations of desert rodents (Heteromyidae). *Animal Behaviour* 45: 263–87.

Randall, John E. 2005. A review of mimicry in marine fishes. *Zoological Studies* 44: 299–328.

Rao, K. Ranga. 2001. Crustacean pigmentary-effector hormones: Chemistry and functions of RPCH, PDH, and related peptides. *American Zoologist* 41: 364–79.

Rapaport, L. G., and G. R. Brown. 2008. Social influences on foraging behavior in young nonhuman primates: Learning what, where, and how to eat. *Evolutionary Anthropology: Issues, News, and Reviews* 17: 189–201.

Rauchenberger, Robert. 2003. Attentional capture by auto- and allo-cues. *Psychonomic Bulletin and Review* 10: 814–42.

Rausch, Robert L. 1953. On the status of some Arctic mammals. *Arctic* 6: 91–148.

———. 1963. Geographic variation in size in North American brown bears, *Ursus arctos* L., as indicated by condylobasal length. *Canadian Journal of Zoology* 41: 33–45.

Redfield, Alfred C. 1930. The absorption spectra of some bloods and solutions containing hemocyanin. *Biological Bulletin* 58: 150–75.

Reichman, O. Jim. 1975. Relation of desert rodent diets to available resources. *Journal of Mammalogy* 56: 731–51.

Reid, P. J., and S. J. Shettleworth. 1992. Detection of cryptic prey: Search image or search rate? *Journal of Experimental Psychology: Animal Behavior Processes* 18: 273–286.

Richards, Owain W. 1961. An introduction to the study of polymorphism in insects. In *Insect Polymorphism: Symposium of the Royal Entomological Society of London,* edited by John S. Kennedy, 2–10. London: Royal Entomological Society

Ridley, Mark. 2004. *Evolution*. 3rd ed. Oxford: Wiley-Blackwell.

Riley, Donald A. 1968. *Discrimination Learning*. Boston: Allyn & Bacon.

Ritland, K., C. Newton, and H. D. Marshall. 2001. Inheritance and population structure of the white-phased "Kermode" black bear. *Current Biology* 11: 1468–72.

Roberts, A. C., T. W. Robbins, and B. J. Everitt. 1988. The effects of intradimensional and extradimensional shifts on visual discrimination learning in humans and non-human primates. *Quarterly Journal of Experimental Psychology* 40B: 321–41.

Robertson, J. M., K. Hoversten, M. Gründler, T. J. Poorten, D. K. Hews, and E. B. Rosenblum. 2011. Colonization of novel White Sands habitat is associated with changes in lizard anti-predator behaviour. *Biological Journal of the Linnean Society* 103: 657–67.

Robertson, J.M., and E.B. Rosenblum. 2009. Rapid divergence of social signal coloration across the White Sands ecotone for three lizard species under strong natural selection. *Biological Journal of the Linnean Society* 98: 243–55.

Robinson, Corinne R. 1921. *My Brother, Theodore Roosevelt.* New York: Charles Scribner's Sons.

Robinson, Michael H. 1969. Defenses against visually hunting predators. *Evolutionary Biology* 3: 225–59.

———. 1970. Insect anti-predator adaptations and the behavior of predatory primates. *Actas del IV Congresso Latinamericano de Zoologia* II: 811–36.

———. 1979. Informational complexity in tropical rain-forest habitats and the origins of intelligence. *Actas del IV Congresso Latinamericano de Zoologia* I: 147–68.

———. 1981. A stick is a stick and not worth eating: On the definition of mimicry. *Biological Journal of the Linnean Society* 16: 15–20.

———. 1990. Predator-prey interactions, informational complexity, and the origins of intelligence. In *Insect Defenses: Adaptive Mechanisms and Strategies of Prey and Predators,* edited by D.L. Evans and J.O. Schmidt, 129–49. Albany: State University of New York Press.

———. 1991. Niko Tinbergen, comparative studies and evolution. In *The Tinbergen Legacy,* edited by M.S. Dawkins, T. Halliday, and R. Dawkins, 100–28. New York: Chapman and Hall.

Röll, Beate. 2001. Retina of Bouton's skink (Reptilia, Scincidae): Visual cells, fovea, and ecological constraints. *Journal of Comparative Neurology* 436: 487–96.

Römpler, H., N. Rohland, C. Lalueza-Fox, E. Willerslev, T. Kuznetsova, G. Rabeder, J. Bertranpetit, T. Schöneberg, and M. Hofreiter. 2006. Nuclear gene indicates coat-color polymorphism in mammoths. *Science* 313: 62.

Roosevelt, Theodore. 1910. *African Game Trails: An Account of the African Wanderings of an American Hunter-Naturalist.* New York: Charles Scribner's Sons.

———. 1911a. On the revealing and concealing coloration of birds: An unpublished letter by Theodore Roosevelt written February 2, 1911. Reprinted by Charles A. Kofoid, 1924. *The Condor* 26: 94–98.

———. 1911b. Revealing and concealing coloration in birds and mammals. *Bulletin of the American Museum of Natural History* 30: 119–230.

———. 1918a. Common sense and animal coloration. *The American Museum Journal* 18: 211–18.

———. 1918b. My life as a naturalist. *The American Museum Journal* 18: 321–50.

Roosevelt, T. and E. Heller. 1914. *Life-histories of African Game Animals.* Vol. 1. New York: Charles Scribner's Sons.

Rosenblum, Erica B. 2006. Convergent evolution and divergent selection: Lizards at the White Sands ecotone. *The American Naturalist* 167: 1–15.

———. 2008. Preference for local mates in a recently diverged population of the lesser earless lizard (*Holbrookia maculata*) at White Sands. *Journal of Herpetology* 42: 572–83.

Rosenblum, E. B., and L. J. Harmon. 2010. "Same same but different": Replicated ecological speciation at White Sands. *Evolution* 65: 946–60.

Rosenblum, E. B., H. E. Hoekstra, and M. W. Nachman. 2004. Adaptive reptile color variation and the evolution of the *MC1R* gene. *Evolution* 58: 1794–1808.

Rosenblum, E. B., H. Römpler, T. Schöneberg, and H. E. Hoekstra. 2010. Molecular and functional basis of phenotypic convergence in white lizards at White Sands. *Proceedings of the National Academy of Sciences* 107: 2113–17.

Rosenthal, Gil G. 2007. Spatiotemporal dimensions of visual signals in animal communication. *Annual Review of Ecology, Evolution, and Systematics* 38: 155–78.

Rothschild, Miriam. 1975. Remarks on carotenoids in the evolution of signals. In *Coevolution of Animals and Plants,* edited by L. E. Gilbert and P. H. Raven, 20–51. Austin: University of Texas Press.

Rowell, C. H. Fraser. 1971. The variable coloration of the acridoid grasshoppers. *Advances in Insect Physiology* 8: 145–98.

Rowland, Hannah M. 2009. From Abbott Thayer to the present day: What have we learned about the function of countershading? *Philosophical Transactions of the Royal Society of London* B 364: 519–27.

———. 2011. The history, theory and evidence for a cryptic function of countershading. In *Animal Camouflage: Mechanisms and Function,* edited by M. Stevens and S. Merilaita, 53–72. Cambridge, UK: Cambridge University Press.

Rowland, H. M., I. C. Cuthill, I. F. Harvey, M. P. Speed, and G. D. Ruxton. 2008. Can't tell the caterpillars from the trees: Countershading enhances survival in a woodland. *Proceedings of the Royal Society* B 275: 2539–45.

Rowland, H. M., M. P. Speed, G. D. Ruxton, M. Edmunds, M. Stevens, and I. F. Harvey. 2007. Countershading enhances cryptic protection: An experiment with wild birds and artificial prey. *Animal Behaviour* 74: 1249–58.

Ruban, Alexander V. 2010. Identification of carotenoids in photosynthetic proteins: Xanthophylls of the light harvesting antenna. In *Carotenoids:*

Physical, Chemical and Biological Functions and Properties, edited by John T. Landrum, 113–36. New York: CRC Press.

Rudge, David W. 2005. The beauty of Kettlewell's classic experimental demonstration of natural selection. BioScience 55: 369–75.

Ruthven, Alexander G. 1907. A collection of reptiles and amphibians from southern New Mexico and Arizona. *Bulletin of the American Museum of Natural History* 23: 483–603.

Ruxton, Graeme D. 2011. Evidence for camouflage involving senses other than vision. In *Animal Camouflage: Mechanisms and Function,* edited by M. Stevens and S. Merilaita, 330–50. Cambridge, UK: Cambridge University Press.

Ruxton, G.D., T.N. Sherratt, and M.P. Speed. 2004. *Avoiding Attack: The Evolutionary Ecology of Crypsis, Warning Signals, and Mimicry.* Oxford: Oxford University Press.

Ruxton, G.D., M.P. Speed, and D.J. Kelly. 2004. What, if anything, is the adaptive function of countershading? *Animal Behaviour* 68: 445–51.

Ryan, James M. 1986. Comparative morphology and evolution of cheek pouches in rodents. *Journal of Morphology* 190: 27–41.

Saccheri, I.J., F. Rousset, P.C. Watts, P.M. Brakefield, and L.M. Cook. 2008. Selection and gene flow on a diminishing cline of melanic peppered moths. *Proceedings of the National Academy of Sciences* 105: 16212–17.

Sage, John H. 1897. Fourteenth Congress of the American Ornithologists' Union. *The Auk* 14: 82–86.

Salvini-Plawen, L.V., and E. Mayr. 1977. On the evolution of photoreceptors and eyes. *Evolutionary Biology* 10: 207–63.

Sandoval, Cristina P. 1994a. Differential visual predation on morphs of *Timema cristinae* (Phasmatodeae: Timemidae) and its consequences for host range. *Biological Journal of the Linnean Society* 52: 341–56.

———. 1994b. The effects of the relative geographic scales of gene flow and selection on morph frequencies in the walking-stick *Timema cristinae. Evolution* 48: 1866–79.

Sandoval, C.P., and B.J. Crespi. 2008. Adaptive evolution of cryptic coloration: The shape of host plants and dorsal stripes in *Timema* walking-sticks. *Biological Journal of the Linnean Society* 94: 1–5.

Sargent, Theodore D. 1976. *Legion of Night: The Underwing Moths.* Amherst: University of Massachusetts Press.

———. 1981. Antipredator adaptations of underwing moths. In *Foraging Behavior: Ecological, Ethological and Psychological Approaches,* edited by A.C. Kamil and T.D. Sargent, 259–87. New York: Garland STPM Press.

Savigny, Jules César de. 1827. *Description de l'Égypte ou Recueil des Observations et des Recherches qui ont été Faites en Égypte Pendant l'Expédition de L'Armée Française.* Vol. 22, plate 2, 439–41. 2nd ed. Paris: C. L. F. Panckoucke.

Schiel, N., and L. Huber. 2006. Social influences on the development of foraging behavior in free-living common marmosets (*Callithrix jacchus*). *American Journal of Primatology* 68: 1–11.

Schiel, N., A. Souto, L. Huber, and B. M. Bezerra. 2010. Hunting strategies in wild common marmosets are prey and age dependent. *American Journal of Primatology* 72: 1039–46.

Schmidly, David J. 1977. *The Mammals of Trans-Pecos Texas: Including Big Bend National Park and Guadalupe Mountains National Park.* College Station: Texas A&M University Press.

Schweickert, Richard. 1993. Information, time and the structure of mental events—a 25-year review. In *Attention and Performance,* edited by D. E. Meyer and S. Kornblum, 535–66. Cambridge, MA: MIT Press.

Scott, George T. 1965. Physiology and pharmacology of color change in the sand flounder, *Scopthalamus aquosus. Limnology and Oceanography* 10, Supplement: R230–46.

Seago, A. E., P. Brady, J.-P. Vigneron, and T. D. Schultz. 2009. Gold bugs and beyond: A review of iridescence and structural colour mechanisms in beetles. *Journal of the Royal Society Interface* 6: S165–84.

Searcy, W. A., and S. Nowicki. 2005. *The Evolution of Animal Communication: Reliability and Deception in Signaling Systems.* Princeton: Princeton University Press.

Seligy, Verner L. 1972. Ommochrome pigments of spiders. *Comparative Biochemistry and Phyisiology* 42A: 699–709.

Shawkey, M. D., N. I. Morehouse, and P. Vukusic. 2009. A protean palette: Colour materials and mixing in birds and butterflies. *Journal of the Royal Society Interface* 6: S221–31.

Shields, Lora M. 1956. Zonation of vegetation within the Tularosa Basin, New Mexico. *The Southwest Naturalist* 1: 49–68.

Sibley, David Allen. 2003. *The Sibley Field Guide to Birds of Eastern North America.* New York: Alfred A. Knopf.

Sillman, A. J., J. K. Carver, and E. R. Loew. 1999. The photoreceptors and visual pigments in the retina of a boid snake, the ball python (*Python regius*). *Journal of Experimental Biology* 202: 1931–38.

Sillman, A. J., V. I. Govardovskii, P. Röhlich, J. A. Southard and E. R. Loew. 1997. The photoreceptors and visual pigments of the garter snake (*Thamnophis sirtalis*): A microspectrophotometric, scanning electron microscopic, and immunocytochemical study. *Journal of Comparative Physiology* A 181: 89–101.

Simon, Hilda. 1971. *The Splendor of Iridescence: Structural Colors in the Animal World.* New York: Dodd, Mead and Company.

Simons, Daniel J. 2000. Attentional capture and inattentional blindness. *Trends in Cognitive Sciences* 4: 147–55.

Simpson, S.J., and G.A. Sword. 2009. Phase polyphenism in locusts: Mechanisms, population consequences, adaptive significance and evolution. In *Phenotypic Plasticity in Insects,* edited by D.W. Whitman and T.N. Ananthakrishnan, 147–89. Enfield, NJ: Science Press.

Simpson, S.J., G.A. Sword, and N. Lo. 2011. Polyphenism in insects. *Current Biology* 21: R738–49.

Sinervo, B., and R. Calsbeek. 2006. The developmental, physiological, neural and genetical causes and consequences of frequency-dependent selection in the wild. *Annual Review of Ecology, Evolution and Systematics* 37: 581–610.

Slominski, A., D.J. Tobin, S. Shibahara, and J. Wortsman. 2004. Melanin pigmentation in mammalian skin and its hormonal regulation. *Physiological Reviews* 84: 1155–1228.

Smith, Hobart M. 1943. The White Sands earless lizard. *Zoological Series of the Field Museum of Natural History* 24: 339–44.

Snowden, R., P. Thompson, and T. Troscianko. 2006. *Basic Vision: An Introduction to Visual Perception.* Oxford: Oxford University Press.

Souto, A., B.M. Bezerra, N. Schiel, and L. Huber. 2007. Saltatory search in free-living *Callithrix jacchus:* Environmental and age influences. *International Journal of Primatology* 28: 881–93.

Speiser, D.I., and S. Johnsen. 2008. Comparative morphology of the concave mirror eyes of scallops (Pectinoidea). *American Malacological Bulletin* 26: 27–33.

Spence, Kenneth W. 1936. The nature of discrimination learning in animals. *Psychological Review* 43: 427–49.

Stebbins, Robert C. 1966. *A Field Guide to Western Reptiles and Amphibians: Field Marks of All Species in Western North America.* Boston: Houghton Mifflin.

Stein, Barbara R. 2001. *On Her Own Terms: Annie Montague Alexander and the Rise of Science in the American West.* Berkeley: University of California Press.

Stein, Gertrude. 1984. *Picasso.* New York: Dover Publications.

Stellar, Eliot. 1960. The marmoset as a laboratory animal: Maintenance, general observations of behavior, and simple learning. *Journal of Comparative and Physiological Psychology* 53: 1–10.

Sterling, Keir B. 1977. *Last of the Naturalists: The Career of C. Hart Merriam.* New York: Arno Press.

Stevens, Martin. 2007. Predator perception and the interrelation between different forms of protective coloration. *Proceedings of the Royal Society* B 274: 1457–64.

Stevens, M., and I.C. Cuthill. 2006. Disruptive coloration, crypsis and edge detection in early visual processing. *Proceedings of the Royal Society* B 273: 2141–47.

Stevens, M., and S. Merilaita, eds. 2011. *Animal Camouflage: Mechanisms and Function*. Cambridge, UK: Cambridge University Press.

Stevens, M., I.S. Winney, A. Cantor, and J. Graham. 2009. Outline and surface disruption in animal camouflage. *Proceedings of the Royal Society* B 276: 781–86.

Steward, Richard C. 1977. Industrial and non-industrial melanism in the peppered moth, *Biston betularia* (L.). *Ecological Entomology* 2: 231–43.

Strathern, Paul. 2008. *Napoleon in Egypt*. New York: Random House.

Stroud, Clyde P. 1949. A white spade-foot toad from the New Mexico White Sands. *Copeia* 1949: 232.

———. 1950. A survey of the insects of White Sands National Monument, Tularosa Basin, New Mexico. *The American Midland Naturalist* 44: 659–77.

Stroud, C.P. and H.F. Strohecker. 1949. Notes on White Sands Gryllacrididae (Orthoptera). *Proceedings of the Entomological Society of Washington* 51: 125–26.

Sugumaran, Manickam. 2002. Comparative biochemistry of eumelanogenesis and the protective roles of phenoloxidase and melanin in insects. *Pigment Cell Research* 15: 2–9.

Sunny, M.M., and A. von Mühlenen. 2011. Motion onset does not capture attention when subsequent motion is "smooth." *Psychonomic Bulletin and Review* 18: 1050–56.

Sussman, R.W., and W.G. Kinzey. 1984. The ecological role of the Callitrichidae: A review. *American Journal of Physical Anthropology* 64: 419–49.

Sutherland, N.S., and N.J. Mackintosh. 1971. *Mechanisms of Animal Discrimination Learning*. New York: Academic Press.

Sykes, Christopher. 1994. *No Ordinary Genius: The Illustrated Richard Feynman*. New York: W.W. Norton & Company.

Tarvin, K.A., and G.E. Woolfenden. 1999. Blue Jay (*Cyanocitta cristata*). In *The Birds of North America*, no. 469, edited by A. Poole and F. Gill, 1–32. Philadelphia: Birds of North America.

Thayer, Abbott H. 1896a. The law which underlies protective coloration. *The Auk* 13: 124–29.

———. 1896b. Further remarks on the law which underlies protective coloration. *The Auk* 13: 318–20.

———. 1902. The meaning of the white under sides of animals. *Nature* 65: 596–97.

Thayer, A. H., and G. H. Thayer. 1919. *Exhibition of Pictures Illustrating Protective Coloration in Nature and Concerned with the Origination of Camouflage in War.* Chicago: Art Institute of Chicago.

Thayer, Gerald H. 1909. *Concealing-Coloration in the Animal Kingdom.* New York: The MacMillan Co.

Théry, Marc. 2007. Colours of background reflected light and of the prey's eye affect adaptive coloration in female crab spiders. *Animal Behaviour* 73: 797–804.

Théry, M., and J. Casas. 2002. Predator and prey views of spider camouflage. *Nature* 415: 133.

Théry, M., M. Debut, D. Gomez, and J. Casas. 2004. Specific color sensitivities of prey and predator explain camouflage in different visual systems. *Behavioral Ecology* 16: 25–29.

Théry, M. and D. Gomez. 2010. Insect colours and visual appearance in the eyes of their predators. In *Advances in Insect Physiology*, vol. 38, edited by J. Casas and S. J. Simpson, 267–353. Burlington, MA: Academic Press.

Théry, M., T. C. Insausti, J. Defrize, and J. Cases. 2011. The multiple disguise of spiders. In *Animal Camouflage: Mechanisms and Function,* edited by M. Stevens and S. Merilaita, 254–74. Cambridge, UK: Cambridge University Press.

Thomas, Keith. 1983. *Man and the Natural World: A History of the Modern Sensibility.* New York: Pantheon.

Thomson, J. Arthur. 1914. *The Wonder of Life.* New York: Henry Holt and Company.

Thoreau, Henry D. 1854. *Walden; or Life in the Woods.* Boston: Ticknor and Fields.

Tinbergen, Luuk. 1960. The natural control of insects in pine woods. I. Factors influencing the intensity of predation by songbirds. *Archives Néerlandaises de Zoologie* 13: 265–343.

Tinbergen, Niko. 1958. *Curious Naturalists.* New York: Basic Books.

———. 1965. Behaviour and natural selection. In *Ideas in Modern Biology,* edited by J. A. Moore, 521–42. New York: Natural History Press.

Tinbergen, N., M. Impekoven, and D. Franck. 1967. An experiment on spacing-out as a defense against predation. *Behaviour* 28: 307–21.

Travis, Anthony S. 2010. Raphael Meldola and the nineteenth-century neo-Darwinians. *Journal for General Philosophy of Science* 41: 143–72.

Treisman, A. M., and G. Gelade. 1980. A feature-integration theory of attention. *Cognitive Psychology* 12: 97–136.

Tristram, Henry B. 1859. On the ornithology of Northern Africa. Part III. The Sahara (continued). *Ibis* 1: 415–35.

———. 1860. *The Great Sahara: Wanderings South of the Atlas Mountains.* London: John Murray.

True, John R. 2003. Insect melanism: The molecules matter. *Trends in Ecology and Evolution* 18: 640–47.

Trullas, S. C., J. H. van Wyk, and J. R. Spotila. 2007. Thermal melanism in ectotherms. *Journal of Thermal Biology* 32: 235–45.

Tucker, Graham M. 1991. Apostatic selection by song thrushes (*Turdus philomelos*) feeding on the snail *Cepaea hortensis. Biological Journal of the Linnean Society* 43: 149–56.

Turatto, M., and G. Galfano. 2000. Color, form and luminance capture attention in visual search. *Vision Research* 40: 1639–43.

Tutt, James W. 1896. *British Moths.* London: George Routledge and Sons, Limited.

Uy, J. A. C., R. G. Moyle, and C. E. Filardi. 2009. Plumage and song differences mediate species recognition between incipient flycatcher species of the Solomon Islands. *Evolution* 63: 153–64.

Uy, J. A. C., R. G. Moyle, C. E. Filardi, and Z. A. Cheviron. 2009. Difference in plumage color used in species recognition between incipient species is linked to a single amino acid substitution in the melanocortin-1 receptor. *The American Naturalist* 174: 244–54.

van't Hof, A. E., and I. J. Saccheri. 2010. Industrial melanism in the peppered moth is not associated with genetic variation in canonical melanisation gene candidates. *PLoS ONE* 5: e10889.

Vevers, Gwynne. 1982. *The Colours of Animals.* London: Edward Arnold.

von Uexküll, Jakob. 1934. *Streifzüge durch die Umwelten von Tieren und Menschen.* Berlin: Springer-Verlag.

Vorobyev, M., and D. Osorio. 1998. Receptor noise as a determinant of colour thresholds. *Proceedings of the Royal Society* B 265: 351–58.

Vuillaume, Monique. 1969. *Les Pigments des Invertébrés: Biochimie et Biologie des Colorations.* Paris: Masson et Cie.

Walker, Ernest P. 1975. *Mammals of the World.* Vol. 2. 3rd ed. Baltimore, MD: Johns Hopkins University Press.

Wallace, Alfred Russel. 1855. On the law which has regulated the introduction of new species. *Annals and Magazine of Natural History* 2d. ser. 16: 184–96.

———. 1858. On the Tendency of Varieties to Depart Indefinitely from the Original Type. Letter to Charles Darwin, February 1858. Accessed February 28, 2012. http://people.wku.edu/charles.smith/wallace/S043.htm.

———. 1865. On the phenomena of variation and geographical distribution as illustrated by the Papilionidae of the Malayan region. *Transactions of the Linnean Society of London* 25: 1–71.

———. 1866. Natural selection. *The Athenaeum*, December 1, 1866: 716–17.

———. 1867. Mimicry, and other protective resemblances among animals. *Westminster Review* 88: 1–43.

———. 1869. *The Malay Archipelago, The Land of the Orang-utan, and the Bird of Paradise. A Narrative of Travel, with Studies of Man and Nature.* London: MacMillan and Company.

———. 1891. *Darwinism: An Exposition of the Theory of Natural Selection with Some of Its Applications.* London: MacMillan and Company.

Walls, Gordon L. 1967. *The Vertebrate Eye and its Adaptive Radiation.* New York: Hafner Publishing.

Ward, M.N., A.M. Churcher, K.J. Dick, C.R.J. Laver, G.L. Owens, M.D. Polack, P.R. Ward, F. Breden, and J.S. Taylor. 2008. The molecular basis of color vision in a colorful fish: Four long wave-sensitive (LWS) opsins in guppies (*Poecilia reticulata*) are defined by amino acid substitutions at key functional sites. *BMC Evolutionary Biology* 8: 210.

Warrant, E., and D.-E. Nilsson, eds. 2006. *Invertebrate Vision.* Cambridge, UK: Cambridge University Press.

Watt, Ward B. 1968. Adaptive significance of pigment polymorphisms in colias butterflies. I. Variation of melanin pigment in relation to thermoregulation. *Evolution* 22: 437–58.

Weiner, Jonathan. 1994. *Beak of the Finch: A Story of Evolution in Our Time.* New York: Alfred A. Knopf.

Weismann, August. 1882. *Studies in the Theory of Descent.* Vols. 1 and 2. Translated and edited by Raphael Meldola. London: Sampson Low, Marston, Searle, & Rivington.

Wells, Jonathan. 1999. Second thoughts about peppered moths: This classical story of evolution by natural selection needs revising. *The Scientist* 13: 13.

Wente, W.H. and J.B. Phillips. 2003. Fixed green and brown color morphs and a novel color-changing morph of the Pacific tree frog *Hyla regilla. The American Naturalist* 162: 461–73.

White, Nelson C. 1951. *Abbott H. Thayer, Painter and Naturalist.* Hartford: Connecticut Printers.

Whiteley, D.A.A., D.F. Owen, and D.A.S. Smith. 1997. Massive polymorphism and natural selection in *Donacilla cornea* (Poli, 1791) (Bivalvia: Mesodesmatidae). *Biological Journal of the Linnean Society* 62: 475–94.

Whitford, Walter G. 2002. *Ecology of Desert Systems.* San Diego: Academic Press.

Whitman, D.W., and T.N. Ananthakrishnan, eds. 2009. *Phenotypic Plasticity in Insects.* Enfield: Science Press.

Whittaker, R.J., and J.M. Fernández-Palacios. 2007. *Island Biogeography: Ecology, Evolution, and Conservation.* 2nd Ed. Oxford: Oxford University Press.

Wickler, Wolfgang. 1968. *Mimicry in Plants and Animals.* New York: McGraw-Hill.

Wilcox, R. Stimson. 1995. Ripple communication in aquatic and semiaquatic insects. *Ecoscience* 2: 109–15.

Williams, Daniel F. 1978. Systematics and ecogeographic variation of the Apache pocket mouse (Rodentia: Heteromyidae). *Bulletin of Carnegie Museum of Natural History* 10: 1–57.

Williams, David. 1991. Jean Louis Rodelphe Agassiz. In *The Great Naturalists,* edited by Robert Huxley, 261–66. New York: Thames and Hudson.

Wilson, D.E., and S. Ruff, eds. 1999. *The Smithsonian Book of North American Mammals.* Washington: Smithsonian Institution Press.

Wilson, E.O., and B. Hölldobler. 1986. Ecology and behavior of the Neotropical cryptobiotic ant *Basiceros manni* (Hymenoptera: Formicidae: Basicerotini). *Insectes Sociaux* 33: 70–84.

Wiltshire, Edward P. 1979. Obituary of Bernard Kettlewell. *Proceedings of the British Entomological and Natural History Society* 12: 101–3.

Winsor, Mary P. 1991. *Reading the Shape of Nature: Comparative Zoology at the Agassiz Museum.* Chicago: University of Chicago Press.

Wolfe, Jeremy M. 1994. Visual search in continuous, naturalistic stimuli. *Vision Research* 34: 1187–95.

———. 1998. Visual search. In *Attention,* edited by Harold Pashler, 13–73. Hove: Psychology Press.

———. 2007. Guided Search 4.0: Current progress with a model of visual search. In Wayne D. Gray, ed., *Integrated Models of Cognitive Systems,* 99–119. New York: Oxford University Press.

Wolfe, J.M., A. Oliva, T.S. Horowitz, S.J. Butcher, and A. Bompas. 2002. Segmentation of objects from backgrounds in visual search tasks. *Vision Research* 42: 2985–3004.

Worthy, T.H., and R.N. Holdaway. 2002. *The Lost World of the Moa: Prehistoric Life of New Zealand.* Bloomington: Indiana University Press.

Wright, Patrick. 2005. Cubist slugs. *London Review of Books* 27: 16–20.

Wright, Sewall. 1978. *Variability within and among Natural Populations.* Vol 4 of *Evolution and the Genetics of Populations.* Chicago: University of Chicago Press.

Yaar, M., and H.-Y. Park. 2012. Melanocytes: A window into the nervous system. *Journal of Investigative Dermatology* 132: 835–45.

Yoshioka, S., and S. Kinoshita. 2002. Effect of macroscopic structure in iridescent color of the peacock feathers. *Forma* 17: 169–81.

Young, R. E., and C. F. E. Roper. 1976. Intensity regulation of bioluminescence during countershading in living midwater animals. *Fishery Bulletin* 75: 239–52.

Zeise, L., M. R. Chedekel, and T. B. Fitzpatrick, eds. 1995. *Melanin: Its Role in Human Photoprotection.* Overland Park: Valdenmar Publishing.

Zentall, Thomas R. 2005. Selective and divided attention in animals. *Behavioural Processes* 69: 1–15.

Zentall T. R., Riley, D. A. 2000. Selective attention in animal discrimination learning. *Journal of General Psychology* 127: 45–66.

Zhang, F., S. L. Kearns, P. J. Orr, M. J. Benton, Z. Zhou, D. Johnson, X. Xu, and X. Wang. 2010. Fossilized melanosomes and the colour of Cretaceous dinosaurs and birds. *Nature* 463: 1075–78.

Zylinski, S., and D. Osorio. 2011. What can camouflage tell us about non-human visual perception? A case study of multiple cue use in cuttlefish (*Sepia* spp.). In *Animal Camouflage: Mechanisms and Function,* edited by M. Stevens and S. Merilaita, 164–85. Cambridge, UK: Cambridge University Press.

Acknowledgments

A REMARKABLE IMAGE inspired our work on this book. Kawika Chetron photographed a kelp gunnel on December 4, 2005, while diving in Monterey Bay. Kawika studied at Harvard University and at the Stanford University School of Engineering and spent his free time photographing off the coast of Northern California. In March 2007, Kawika was lost at sea off the coast of Mendocino. We thank his mother, Ibbie White Al-Shamma, for permission to include his photograph in this book.

We express our appreciation to many libraries and archives. For permission to reproduce published or unpublished materials, we thank: the President and Fellows of Wolfson College and the Bodleian Library Collections of the University of Oxford for permission to cite E. B. Ford's July 19, 1948, letter to Bernard Kettlewell; the Musee National d'Histoire Naturelle in Paris for permission to photograph the female peppered moth collected by J. F. Fallou in 1859 and donated to the entomology collections in 1895; and the Hope Entomological Collections at the Oxford University Museum of Natural History for permission to photograph peppered moths from the William Jones Collection and to James Hogan, Darren J. Mann, and D. J. Rogers for providing us with access to the collection.

For use of their beautiful photographs, we thank: Dr. Matthias Borer, Curator in the Department of Entomology at the Natural History Museum, Neuchâtel, Switzerland, for use of his photographs of desert mantids; Bert Hölldobler for use of the photograph of the *Basiceros* ants from *The Ants* by Bert Hölldobler and Edward O. Wilson, Belknap Press of Harvard University Press; Joel Sartore for use of his photograph of the blue jay; David Bustos,

National Park Service, White Sands National Monument, for use of his photograph of the White Sands camel cricket; Michael Nachman for providing us with his photographs of color variations of rock pocket mice; Erica Bree Rosenblum for use of her photograph of little striped whiptails; Simone Des Roches for use of her photograph of lesser earless lizards; Martin Stevens for use of images of his pepper moth models; Paul V. Loiselle for use of his photograph of a male guppy; Wade Sherbrooke for putting us in touch with Jeff Judd at http://projectphrynosoma.com and Jeff for use of his photos of *Phrynosoma hernandesi*; Gregory Sword for use of his photograph of the locust *Schistocerca gregaria*; and Krushnamegh Kunte, Ph.D., National Center for Biological Sciences (NCBS), Tata Institute of Fundamental Research (TIFR), India, for use of his *Papilio memnon* image.

For access to field sites, we are especially grateful to the Southern Islands Area of the New Zealand Department of Conservation, the iwi authority, Te Rünanga o Ngäi Tahu, and the Whenua Hou Management Committee for permission to conduct research on Whenua Hou Codfish Island. We also thank Kay E. Holcamp, Professor, Department of Zoology, Michigan State University, and her student, Sarah Benson-Amram, for hosting us at their field site in the Masai Mara National Reserve, Kenya.

Several local Nebraska institutions aided our project immensely. We acknowledge the Lincoln Children's Zoo for permission to photograph their flamingos. The excellent libraries at the University of Nebraska were an essential aid throughout this project. We are beholden to our home units, the University of Nebraska State Museum and the School of Biological Sciences, for giving us the time and encouragement to pursue this project. We were supported during the writing and research for this book by several federal grants: (Diamond) NSF Grants DRL-0229294,

DRL-0632175, DRL-0715287, and DRL-1010889; (Bond) NSF Grants IBN-9631044, IBN-99746669, and IBN-023441; and NIMH Grants MH-068426, MH-083325, and MH-069893.

It was a great pleasure to work with the talented scientific illustrator at the University of Nebraska State Museum, Angie Fox, who provided the illustrations in this book. We are also grateful to the Division of Entomology at the University of Nebraska State Museum for permission to use Angie's photograph of variation within a single beetle species, *Cyclocephala sexpunctata*.

We were convinced to write this book by Ann Downer-Hazell, to whom we are indebted for her patience and persistence. Courtney McCusker provided diligent editing and review, for which we are very grateful. Our good friend, Jeannie Allen, from NASA Goddard Space Flight Center, helped locate Landsat images. We thank Lauren K. Esdaile for her assistance, and two anonymous reviewers for their valuable suggestions. Our editor, Michael Fisher, at Harvard University Press, provided essential encouragement and helpful guidance.

We were inspired and assisted by many members of our extended families. Our son, Benjamin, reviewed chapters and gave us valuable editorial advice. Our daughter, Rachel, assisted us in the field and convinced us this project would be worthwhile. Over the years, we appreciated discussions about this work with family members Alon, Josh, Keren, Sarah, Jordie, David, Julian, Erica, Darya, Wiley, Felix, and, more recently, Jack and Faye.

Index

Page numbers in italic refer to figures.